International Perspectives on Aging

Series Editors:
Jason L. Powell
Sheying Chen

For further volumes:
http://www.springer.com/series/8818

Pranitha Maharaj
Editor

Aging and Health in Africa

Springer

Editor
Pranitha Maharaj
School of Built Environment
and Development Studies
University of KwaZulu-Natal
4041 Durban, South Africa

ISBN 978-1-4419-8356-5 ISBN 978-1-4419-8357-2 (eBook)
DOI 10.1007/978-1-4419-8357-2
Springer New York Heidelberg Dordrecht London

Library of Congress Control Number: 2012943041

Printed on acid-free paper

Springer is part of Springer Science+Business Media (www.springer.com)

Preface

Africa is a relatively youthful continent, but projections suggest that in the future older people will constitute a proportionally larger share of the total population. The shift towards an ageing population is largely a result of improvements in food production and distribution, water and sanitation, advances in medical technology, as well as changes in child spacing and family size. The book shows that the ageing of the population in Africa varies considerably by region. The vast majority of the older populations are concentrated in the most populated regions of the continent. However, the elderly constitute a significant share of the total population of the less populated regions of Africa which have the most rapidly ageing populations in the continent. The ageing of the population is likely to have major and far-reaching implications for a continent with widespread poverty, decades of instability and civil strife, changing household structures, a heavy burden of communicable diseases, inadequate health systems, and weak or poorly managed political institutions.

Despite the evident need to understand issues that affect the health situation of the population, the book shows that in many parts of Africa there has not been adequate focus on the older population. As people live longer in Africa, the health profile of the population is likely to change as many chronic illnesses increase with age. As health deteriorates with advancing age, frailty and disability become more frequent and there is an increased demand for health care. The shift in the health profile of the population will have considerable implications for health service provision and resource allocation. Traditionally, in Africa, older people were primarily supported by the family. However, increasing development and urbanization, together with socioeconomic and political changes, have resulted in the weakening of traditional social support networks. As a result, the social realities of ageing as well as the implications for national policies need to be better understood.

This book documents the realities facing the population as they grow old in Africa. In many parts of Africa, the older population live in poverty, which often exacerbates the degenerative effects of ageing. A major challenge is access to health care, especially for those who are poor and living in rural areas where distances to clinics are great and public transport scarce. The AIDS pandemic is also adding to the stress of the older population as they increasingly have to take responsibility for sick and/or dying children and grandchildren. This book is timely—as the number of older people in the population steadily increases there is an obvious need to explore their needs and contribution.

Information that is relevant and specific to the older population and their situation is a high priority area given the dearth of adequate data. Improving the quality and availability of data will assist in improving our understanding of the ageing process in Africa. In Africa the issue of population ageing has often been marginalized. It is therefore of vital importance that ageing issues are given serious consideration in national programmes and policies and governments in Africa take measures to face the challenges posed by population ageing.

Durban, South Africa Pranitha Maharaj

Contents

Acronyms

AGRAD	Ageing with Grace and Dignity
AU-Plan	African Union Policy Framework and Plan of Action on Ageing
CBR	Community based rehabilitation
CDVTA	Community Development Volunteers for Technical Assistance
CSG	Child Support Grand
DRC	Democratic Republic of Congo
FNR	National Retirement Fund
GHS	Ghana Statistical Service
GNP	Gross National Product
GPRS	Growth and Poverty Reduction Strategy II
HIV/AIDS	Human immunodeficiency virus/Acquired immune deficiency syndrome
HSNP	Hunger Safety Net Programme
ICESCR	International Covenant on Economic, Social and Cultural Rights
ICPD	International Conference on Population and Development
ID	Identification card
IFAD	International Fund for Agricultural Development
ILO	International Labor Organization
INDEPTH	International Network for the Demographic Evaluation of Populations and Their Health
IPRES	Institute of Retirement of Senegal
LEAP	Livelihood Employment Against Poverty
MBOSCUDA	Mbororo Social and Cultural Development Association
MCH	Maternal and Child Health
MDG	Millennium development goals
MIPAA	Madrid International Plan of Action on Ageing
NCD	Non-communicable disease
NGO	Nongovernmental Organization
NHIS	National Health Insurance Scheme
NIDS	National Income Dynamics Study

OAP	Old age pension
PHC	Primary Health Care
SAGE	Study on Global AGEing and Adult Health
TB	Tuberculosis
UBOS	Uganda Bureau of Statistics
UN	United Nations
URAA	Uganda Reach the Aged Association
WHO	World Health Organization

Contributors

Farzana Alli Health Economics and HIV/AIDS Research, University of KwaZulu-Natal, Durban, South Africa

Nana Apt Ashesi University College, Berekaso E/R, Ghana

Charles Che Fonchingong Canterbury Christ Church University, Department of Health, Wellbeing and Family, CCCU, Canterbury, UK

Ashley Gresh School of Built Environment and Development Studies, University of KwaZulu-Natal, Durban, South Africa

Nina Hunter School of Built Environment and Development Studies, University of KwaZulu-Natal, Durban, South Africa

Pranitha Maharaj School of Built Environment and Development Studies, University of KwaZulu-Natal, Durban, South Africa

Julian May School of Built Environment and Development Studies, University of KwaZulu-Natal, Durban, South Africa

Chantal Munthree School of Built Environment and Development Studies, University of KwaZulu-Natal, Durban, South Africa

Annet Nankwanga Department of Biochemistry and Sport Science, Makerere University, Kampala, Uganda

Stella Neema Department of Sociology, Makerere University, Kampala, Uganda

Julie Phillips Physiotherapy Department, University of Western Cape, Cape Town, South Africa

Natashya Kristanna Pillay Health Economics and HIV/AIDS Research Division (HEARD), University of KwaZulu-Natal, Durban, South Africa

School of Built Environment and Development Studies, University of KwaZulu-Natal, Durban, South Africa

Ganzamungu Zihindula School of Built Environment and Development Studies, University of KwaZulu-Natal, Durban, South Africa

Chapter 1
Introduction

Pranitha Maharaj

This book explores the connection between ageing and health in Africa. Population ageing—the process by which older people increasingly constitute a proportionally larger share of the total population—is a matter of concern. Improvements in food production and distribution, water and sanitation, advances in medical technology, as well as a number of factors related to child spacing and family size have all led to more people living longer. As outlined in the Madrid International Plan of Action on Ageing in 2002, nations should take a holistic life course approach to ageing to enable older people to maintain their independence, productivity, and remain vital resources for their families, communities, and the economy; good health is crucial to achieve this. Advancing health includes confronting policy and programmatic issues such as health promotion, universal and equal access to health care services, HIV/AIDS, and training of health professionals and older persons. It is important to integrate these ageing issues into programmes and policies to achieve the Millennium Development Goals (MDGs) and to provide commitment to population ageing issues in Africa.

This book argues that the ageing of the population has major and far reaching consequences, not only for the elderly but also the wider society. Population ageing often occurs in tandem with changes in the health profile of the population. In Africa, many countries are already facing a high burden of infectious diseases. As people grow older they are also more likely to experience chronic health problems associated with the ageing process. Population ageing in Africa is occurring in the context of widespread poverty, instability and conflict, changing household structures, a high disease burden, inadequate health systems, and weak or poorly managed political institutions. However, most national governments in Africa have not begun to sufficiently address the issue of how to respond effectively to the

P. Maharaj (✉)
School of Built Environment and Development Studies, University of KwaZulu-Natal, Durban, South Africa
e-mail: Maharajp7@ukzn.ac.za

P. Maharaj (ed.), *Aging and Health in Africa*, International Perspectives on Aging 4,
DOI 10.1007/978-1-4419-8357-2_1, © Springer Science+Business Media New York 2013

challenge of population ageing. This will require a better understanding of the situation of the older population in Africa. This book attempts to fill the gaps that exist by exploring the social realities of population ageing in Africa. It is intended for an interdisciplinary audience of sociologists, demographers, public health professionals, and advocacy groups. The intention is to target both academic and non-academic readers, including policy makers and civil society organisations working on ageing and health issues.

The introductory chapter is organised as follows. First, the chapter will provide an overview of population ageing and focus more specifically on some of the difficulties in studying the health situation of older people in Africa. It will also look at some of the challenges of ageing in the African continent. Finally, it will end by briefly describing the chapters and close with some reflections for future work on this area.

1.1 Overview of Population Ageing in Africa

Over the past few decades, the world has experienced a rapid growth in population. The world population increased from 2.5 billion in 1950 to 7 billion in 2011, and projections suggest that the increase is likely to continue for several more decades. Most of the growth has been occurring in developing countries. The population of Africa is one of the smallest in the world, but is expected to double by 2050 despite a devastating AIDS pandemic. The population of the world is not only growing faster, but it is also becoming considerably older.

According to the United Nations Population Division, during the next 5 decades, the number of persons in the world aged 60 years or older is expected to almost triple, increasing from 672 million people in 2005 to nearly 2 billion by 2050 (United Nations 2007). Although population ageing is more apparent in developed regions, it is increasingly gaining importance in less developed regions. In 2000, almost one-fifth of the population in the more developed regions was aged 60 and over, but only 8% in the less developed regions. By 2050, 1 in every 3 persons in the more developed regions and 1 in every 5 in less developed are projected to be 60 and over (United Nations 2007). By the middle of the century, the developing world is likely to reach the same level in the process of population ageing as the developed world (United Nations 2009). Although the percentage of older persons is higher in the more developed regions, the number of older persons is increasing at a more rapid pace in the less developed regions over a shorter period of time, including Africa.

In 1950, the population aged 60 and above was approximately 12 million in Africa. By 2007, this number had risen to about 50.5 million people (United Nations 2007). According to the World Population Prospects 2006 Revision, the population aged 60 and over in Africa will reach 64.5 million by 2015, which is also the target date for attaining the MDGs. By 2030 projections suggest that there will be 103 million older people, and this will increase to 205 million by 2050. In terms of proportion of the total population, the percentage of the population aged 60 and above grew from 4.9 to 5.3% between 1950 and 2005. Projections suggest an increase from 5.6% in 2015, to 6.8% in 2030, and 10.4% in 2050. The reality is that older people will constitute an increasingly significant share of the population in Africa (United Nations 2007). Population ageing in Africa varies considerably by region. The vast majority of the older populations is concentrated in Northern, Eastern, and Western Africa—the most populated regions of the continent. In contrast, smaller fractions are concentrated in Central and Southern Africa. However, the elderly constitute a significant share of the total population of Northern and Southern Africa. Both these regions have the most rapidly ageing populations in the African continent (Apt 2000).

Most developed countries have had several decades to adjust to this change in the age structure of the population. For example, it took close to a century for the population of Sweden aged 65 and over to rise from 7 to 14% of the total population. In sharp contrast, developing countries are forced to adapt to a new age structure over a shorter period of time with a much larger population base. For example, in Tunisia it took only less than 3 decades for the population aged 65 and over to rise from 7 to 14% of the total population (Kinsella and Gist 1995). Many developing countries are likely to confront population ageing without the economic growth that accompanied the ageing process in many of the developed countries.

Until relatively recently, ageing in Africa has received limited attention. This is partly related to the lack of consensus on the definition of old age. The age of retirement in most developed countries is widely regarded as the beginning of old age. In developed countries, chronological age is given far greater prominence than in developing countries. However, in Africa, chronological age—which is sometimes not even known—has little or no relevance in making sense of the ageing process, apart from the fact that in some countries it is the age at which one begins to receive a pension. In the African context, some argue that this definition is not seen as particularly useful, given that the majority of older people work outside the formal sector, mainly in agriculture, and thus are unlikely to receive retirement benefits. Other social constructions of age are of greater importance in trying to understand the process of ageing in Africa. These social constructions arise from the roles that are assigned to older people or, perhaps more importantly, the loss of roles that accompany physical decline (Heslop and Gorman 2002). Old age is thus seen to begin at that point in life when people, because of physical decline, are no longer able to actively carry out their work and family roles. Functional rather than chronological age is seen as an important indicator of ageing in a rural subsistence context (Roebuck 1979). According to the functional definition, *the aged* refers to those who are infirm, frail, and suffering incapacities to the extent that they are no longer able to fully support or take care of themselves, and who also display the characteristics of being old (Roebuck 1979). However, functional ageing does not always correspond identically with chronological age. For example, it is rightly pointed out that two persons may share the same chronological age but they may differ vastly in their functional capabilities. Thus, people younger than 60 years may be widely regarded as old because they share similar physical characteristics and patterns of morbidity as people over the age of 60 years. It is important to recognise that older people do not form a homogenous group. In addition to having widely varying abilities within the cohort, there is also likely to be considerable diversity by gender, class, ethnicity, and nationality.

Understanding the diversity of population ageing is heavily dependent on the availability of complete and reliable data. The current situation of the older population in many African countries is not well known. Very limited data is available for many countries in Africa because research efforts have been severely hampered by the lack of infrastructure due to decades of social upheaval, armed conflict, and political mismanagement. An important source of health data in many developing countries is the Demographic and Health Survey, but it usually excludes the older population. Over the past decade there have been concerted efforts in some parts of Africa to try to improve the data situation and profile the situation of the older population. The most notable is the WHO Study on Global Ageing (SAGE), conducted in six countries, two of which (Ghana and South Africa) are in Africa. While the paucity of data on ageing in Africa has been identified as one of many challenges, an additional problem is the quality of the data. The poor quality of data in many African countries is reflected in the incomplete recording of vital statistics, particularly with regard to causes of mortality. In addition, another problem includes data that is not always readily available in an easily accessible or user-friendly format, which makes analysis difficult. Moreover, there is often a long time lapse between the collection of the data and the release of the data. Sometimes data is released without sufficient attention to cleaning, and this impacts the quality of the data (Ferreira and Kowal 2006). In many African countries, there is a severe shortage of appropriately trained and skilled personnel to undertake quality research.

In the past decade there has been more emphasis on improving availability and quality of data in order to profile the situation of the older population. Interest in understanding the process of ageing within and across countries is steadily growing. There is increasing recognition that there is a need for more complete and reliable information on the physical, psychological, and socio-economic well-being of older Africans for the formulation and implementation of informed policies.

1.2 Challenges of Healthy Ageing in Africa

Increasingly, it is becoming evident that the health status of older people is changing not only because the population is growing older, but also as a result of shifts in the burden of disease. Africa faces a greater set of health challenges than any other major parts of the world and has a particularly heavy burden of diseases including malaria and tuberculosis; today it remains the epicentre of the AIDS pandemic. Malaria is a major cause of morbidity and mortality in Africa. According to the World Health Organization, estimates in 2006 show that there were 190–330 million malaria episodes, leading to nearly 1 million malaria-related deaths (World Bank 2011). Worldwide, tuberculosis (TB) is another major cause of morbidity and mortality, accounting for nearly 2 million deaths every year. The prevalence of tuberculosis has levelled off but remains highest in Africa (World Bank 2011). Africa is also suffering from a devastating AIDS pandemic. Worldwide, more than 30 million people—two-thirds in sub-Saharan Africa—are living with HIV/AIDS, but the prevalence rate has not changed substantially since 2000. In six countries, life expectancy has fallen since 1970: Democratic Republic of the Congo (DRC), Lesotho, Swaziland, South Africa, Zambia, and Zimbabwe; there are all countries where AIDS prevalence rates still exceed 15% (UNDP 2010). Improved access to antiretroviral treatment has contributed to a decline in mortality since the beginning of the AIDS pandemic. The high disease burden of malaria, tuberculosis, and AIDS may make the older population increasingly vulnerable to illnesses and subsequently early death. It also places tremendous pressure on health systems that are already severely constrained by limited funding, inadequate health infrastructure, shortage of health personnel, and lack of consistent drug supplies.

As more and more people transition from adulthood to old age, they are also more likely to experience health ailments associated with increased longevity which may require medication and sometimes even specialised treatment. Studies suggest that the older populations are more likely to experience malnutrition, chronic physical and mental conditions, hearing and sight difficulties, depression, and dementia (Aboderin 2010). The movement of older people may also be severely diminished because of ill-health. Increasing health ailments may curtail their independence and ability to carry out their normal routine of daily activities. The longer the person lives, the greater the likelihood that he or she will require instrumental support (i.e. help with such tasks as cooking and shopping), as well as financial and emotional support. This is particularly true when he or she is no longer actively employed and begins to experience health ailments that curtail his or her dexterity and ability to carry out tasks necessary for daily survival (Zimmer and Dayton 2003). In this context, older people are also likely to find it difficult to maintain their social relationships and they may find themselves increasingly isolated and alone. As health deteriorates with advancing age, frailty and disability become more frequent and there is also greater dependence on external sources for support. The increasing prevalence and complexity of chronic diseases as the population grows older will increase pressure on health systems in Africa.

A distinct feature of ageing in Africa is that a large proportion of the older population lives in rural areas. In many rural areas in Africa, the elderly constitute the majority of the population. Most often, the number of women outnumbers the men in rural areas. It is projected that by the year 2020 approximately 64% of Africa's elderly will live in areas defined as rural (HelpAge International 2004). In general, rural areas have a higher dependency burden than urban areas: a higher child dependency ratio and a higher old age dependency ratio (United Nations 2009). Increasing urbanisation and migration have led to the movement of young people out of rural areas, and this has meant fewer people to take care of the elderly. Moreover, rural areas are characterised by high levels of poverty, poor housing and transportation systems, and lack of a wide range of basic social services. Poverty and long distances to health facilities limit access to services for the older population.

Despite the associated need for greater health care in later life, older people in developing countries are often faced with health care that is unaffordable. In many instances, older people are required to pay user fees, which they often do not have, in order to access health care. In addition, they have to spend money on transportation costs, walk long distances to get to a health facility, and leave home early in the morning to spend most of the day sitting in a long queue so that they can been seen by a health provider. Not surprisingly, many older people are deterred from seeking the necessary and appropriate medical care. In Africa, poverty is widespread, and often the older sector of the population constitutes the poorest of the poor.

The majority of African nations have made major gains in human development in the past 40 years. Among African countries that have registered substantial progress in improving human development, Ethiopia ranks 11th in progress over time, with Botswana, Benin, and Burkina Faso among the 25 "Top Movers" in the world (UNDP 2010). Advancements in education have been widespread in Africa. The average literacy rate nearly tripled in percentage terms over the past 4 decades, growing from 23% in 1970 to 65% today. In addition, the average life expectancy in sub-Saharan Africa is currently 52 years, up from 44 years in 1970 (UNDP 2010). The proportion of Africans surviving on less than $1.25 a day also declined from 58% in 1990 to 51% in 2005 (World Bank and International Monetary Fund 2010). Though parts of the continent have made substantial progress over the last few years, the majority of the least developed countries are in Africa. The group of least developed countries, according to the United Nations, comprises 49 countries, of which 33 are in Africa, 10 in Asia, 1 in Latin America and the Caribbean, and 5 in Oceania. These countries rank the lowest in terms of socio-economic development. In addition, according to the most recent Human Development Report, the region is home to three countries with a lower human development index today than in 1970: Democratic Republic of the Congo, Zambia, and Zimbabwe (UNDP 2010).

The recent financial crisis is widely regarded as the worst economic downturn in 50 years and led to the global economic recession (World Bank 2011). A recent report suggests that the economic crisis is likely to slow the pace of poverty reduction in Africa and hamper progress toward the MDGs (World Bank and International Monetary Fund 2010). At present, sub-Saharan Africa also has the greatest incidence of multi-dimensionally poor in the world. There are regional variations in poverty levels, with a low of 3% multi-dimensionally poor (multi-dimensional poverty identifies simultaneous deprivations in health, education, and living standards on the household level) in South Africa to a high of 93% in Niger (UNDP 2010). Recent projections suggest that the poverty rate for sub-Saharan Africa will be 38% by 2015, rather than the 36% in the absence of the crisis (World Bank and International Monetary Fund 2010). The continent remains one of the least developed and by far the poorest in the world. More than 50% of the population of Africa is classified as poor, which contrasts sharply with other regions of the world that have made great strides in combating poverty (UNDP 2010). In contrast to countries in East Asia, the absolute number of poor people in sub-Saharan Africa grew from 296 million to 388 million (World Bank and International Monetary Fund 2010). In many African countries, annual economic growth has been minimal or negative and inequalities in living standards are particularly pronounced (UNDP 2010). Economic growth is a major driver in combating poverty and achieving other desired development outcomes (World Bank and International Monetary Fund 2010). A number of factors have contributed to Africa's poor economic situation, including political strife and instability. Political conflict has had long-term effects, including the disintegration of the social fabric; loss of family; disruption of daily life; lack of access to food, shelter, and medical care; the breakdown of basic services; destruction of the local infrastructure; and increased vulnerability to disease (Pederson 2002). It also impacts negatively on an individuals' sense of security and well-being, particularly the elderly. Many elderly find themselves severely restricted due to their limited mobility or physical strength, and without anyone to care for them in times of political instability. In the absence of social welfare systems, older people are particularly vulnerable to high levels of poverty. In most developed countries, old age support comes, to a greater extent, from public or private pensions. However, with a few notable exceptions, the majority of African countries do not offer formal social protection schemes. The elderly, therefore, often have to rely on their own current and accumulated earnings and financial support from family members (Gillian et al. 2000). However, it is also becoming increasingly evident that traditional family support is declining because of changes in family structures as a result of migration and urbanisation as well as other factors.

The challenges posed by rising numbers of older persons in Africa require urgent attention because of widespread poverty, the AIDS pandemic, political instability, and heavy disease burden facing the continent. Most countries in Africa will confront population ageing to some degree over the next few decades, and there is a great window of opportunity for new thinking and careful planning for the future.

1.3 Summary and Organisation of the Volume

The population of Africa is not only growing quickly, but it is also increasingly becoming older in the context of high levels of poverty, low levels of education, lack of social security, political conflict, as well as inadequate health and social services. The ageing of the population is closely linked to the health of the population. Africa is currently confronting a high burden of communicable and infectious diseases including malaria, tuberculosis, and HIV/AIDS. As people live longer, the pattern of diseases is likely to change, increasing medical costs as well as accelerating demand for health care. Health is intimately connected to ageing because good health often means the ability to continue participating in the labour force and achieving a reasonable standard of living, and is crucial for creating and maintaining a productive society (Randel et al. 1999). As people in Africa grow older they increasingly have to cope with widespread poverty without access to social security. With the exception of a few countries in Africa, social security systems are virtually non-existent. The family represents a vital source of support for the older population, but with increasing urbanisation and migration, traditional care systems are on the decline. Population ageing has not been a priority area for national governments given the host of problems facing the continent. More emphasis should be placed on developing policies that mitigate the negative consequences of population ageing in Africa. It is important to recognise that not only is population ageing a challenge for developed regions, but it is also an issue for developing countries.

This volume is organised into three parts. The first part sets the scene for the volume, profiling the demographic and health situation of the elderly in Africa. Part two examines specific country case studies to better illuminate the social realities of ageing in the continent. The case studies use a combination of qualitative and quantitative data to capture the process of ageing and its impact on health in a range of countries. Some of these case studies draw on data collected in countries that have very limited data. The final part considers the policy and programmatic response to ageing in Africa. It looks more specifically at how the response has shifted and what programmes are being implemented to meet the health needs of older men and women in Africa. Together these chapters constitute a useful vantage point from which to undertake further research on the topic.

References

Aboderin, I. (2010). Understanding and advancing the health of older populations in sub-Saharan Africa: policy perspectives and evidence needs. *Public Health Reviews, 32*, 357–376.

Apt, N. (2000). *Rapid urbanization and living arrangements of older persons in Africa*. New York: United Nations Department of Economic and Social Affairs.

Ferreira, M., & Kowal, P. (2006). A minimum data set on ageing and older persons in Sub-Saharan Africa. *African Population Studies, 21*(1), 19–36.

Gillian, C., Turner, J., Bailey, C., & Latulippe, D. (2000). *Social security pensions: Development and reform*. Geneva, Switzerland: International Labour Office.

HelpAge International. (2004). *Ageing issues in Africa: A summary*. London, UK: HelpAge International.

Heslop, A., & Gorman, M. (2002). *Chronic poverty and older people in the developing world, Chronic Poverty Working Paper No. 10*. Manchester: Chronic Poverty Research Centre.

Kinsella, K., & Gist, Y. (1995). *Older workers, retirement, and pensions. A comparative international chartbook*. Washington, DC: U.S. Census Bureau and U.S. National Institute on Aging.

Pederson, D. (2002). Political violence, ethnic conflict, and contemporary wars: broad implications. *Social Science and Medicine, 55*(2), 175–190.

Randel, J., German, T., & Ewing, D. (Eds.). (1999). *The ageing and development report: Poverty independence and the World's older people*. London: Earthscan Publications Limited.

Roebuck, J. (1979). When does old age begin?: The evolution of the English definition. *Journal of Social History, 12*(3), 416–428.

UNDP. (2010). *Human Development Report 2010. The real wealth of nations: Pathways to development*. New York: Palgrave Macmillan.

United Nations. (2007). *World population prospects: The 2006 revision*. New York: United Nations Department of Economic and Social Affairs.

United Nations. (2009). *World population ageing 2009*. New York: United Nations Department of Economic and Social Affairs.

World Bank. (2011). World development indicators 2010. World Bank. http://data.worldbank.org/data-catalog/world-development-indicators/wdi-2010. Accessed 10 Aug 2011.

World Bank and International Monetary Fund. (2010). *Global monitoring report 2010: The MDGs after the crisis*. Washington, DC: IBRD.

Zimmer, Z., & Dayton, A. (2003). *The living arrangements of older adults in Sub-Saharan Africa in a time of AIDS. Policy Research Division Working Papers Series, No. 169*. New York: Population Council.

Chapter 2
Population Ageing in Africa

Natashya Kristanna Pillay and Pranitha Maharaj

2.1 Introduction

Worldwide, the number of persons aged 60 and over has been increasing at an unparalleled rate. In 1980, there were 378 million people aged 60 or above; 3 decades later, this figure doubled to 759 million and by 2050 is projected to rise almost threefold to 2 billion people (United Nations 2010). In almost all of the regions of the world, the older population is growing faster than the total population (United Nations 2009). In particular, the older population in developing countries has a higher speed of growth than in developed countries. Compared with other regions of the world, the population of Africa is growing older faster, at a rate of 2.27% (United Nations 2011). Although the size of the older population in percentage terms is expected to remain small, the absolute number of older persons is expected to increase dramatically over the next few decades.

Africa, like other parts of the world, is undergoing rapid demographic changes, and although the population is largely youthful, the proportion of older persons has increased tremendously over the past few decades. The growth of the ageing population of Africa is accompanied by an increase in the median age of the population, as well as changes in the dependency ratio, resulting in a decline in the proportion of the population composed of children, and an increase in the population aged 60 years and

N.K. Pillay (✉)
Health Economics and HIV/AIDS Research Division (HEARD),
University of KwaZulu-Natal, Durban, South Africa

School of Built Environment and Development Studies,
University of KwaZulu-Natal, Durban, South Africa
e-mail: natashyakpillay@gmail.com

P. Maharaj
School of Built Environment and Development Studies,
University of KwaZulu-Natal, Durban, South Africa
e-mail: Maharajp7@ukzn.ac.za

P. Maharaj (ed.), *Aging and Health in Africa*, International Perspectives on Aging 4,
DOI 10.1007/978-1-4419-8357-2_2, © Springer Science+Business Media New York 2013

over. The change in the age structure of the African population is likely to have far-reaching consequences for the continent. This requires a shift in policy-making and developmental efforts in order to respond to the changes resulting from this demographic phenomenon. In order to achieve this, accurate and reliable data on the ageing population in Africa must be easily and widely accessible to ensure that programmes incorporate current and projected population trends into their planning processes.

These requirements can be assisted by the scope of the following chapter which begins by discussing the context of population ageing in Africa, its features at a continental level, and its determinants. This is followed by an examination of the demographic characteristics of the ageing population in Africa, which includes information on the magnitude and pace of population ageing being experienced by the continent. The next section profiles the absolute size, percentages, and growth rates of each of the five African regions. Thereafter, current and future demographic trends in ageing are explored within each African region. This is followed by a country-level examination of the differentials in population ageing across and within African countries. The chapter concludes with a discussion of the implications of these demographic changes on the African ageing population.

2.2 Population Ageing in Africa

In many parts of Africa, older people are making a valuable contribution to society—from providing care for sick and/or dying children and orphaned grandchildren to providing much needed financial support for the household (HelpAge International 2008). In traditional African societies, the elderly occupied a high status in the community. Relations between the young and the old were marked by respect for the elderly and recognition of their accumulated wisdom, experience, and authority (Rwezaura 1989). With the advent of modernisation, urbanisation, and migration, there has been a marked change in attitudes towards the elderly. The elderly have been largely ignored or excluded. The emphasis that is placed on the younger generation is to some extent justified as Africa is a relatively youthful continent, with more than 40% of its population below the age of 15 years. A youthful African population has meant that the elderly often do not feature in national policies. However, projections for the future suggest that the age structure of Africa is likely to be altered dramatically, and the continent is likely to experience an accelerated rate of growth of numbers of older people in the population than any region in the world, and thus there should be more focus on this group. What remains an issue is the inadequate care and consideration given to the ageing population in development initiatives. This becomes an urgent priority given that Africa has one of the fastest growing rates in the world.

However, an issue for many African countries is the lack of data on a national scale that will assist in making informed decisions; this is especially so for the ageing population (UNFPA 2008a). What can be stated with certainty is that the ageing population in Africa will double in size from 25 million to 52 million people over a span of 25 years (UNFPA 2008a). On a proportional scale, in the year 2000 those

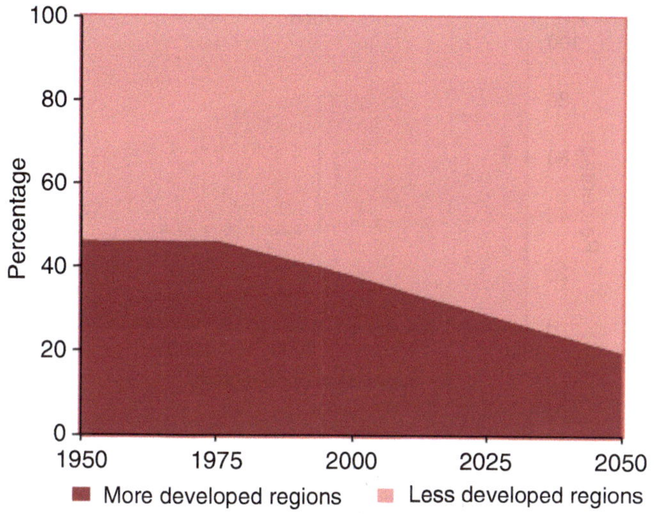

Fig. 2.1 The distribution of the population aged 60 years and over by development regions 1950–2050. *Source*: United Nations (2001)

over 60 years of age constituted 5.1% of the African population and this figure is expected to more than double to 10.4% in 2050 (United Nations 2007). One would assume that the AIDS pandemic would reduce the life expectancy of the population and thus prevent the growth of the older population, but according to Van Dulleman (2006), in the next 3 decades, neither AIDS, nor domestic wars will prevent the number of elderly in most African countries from doubling.

On a proportional scale less, developed regions have had a larger ageing population than more developed regions. As Fig. 2.1 illustrates, during the 1950s, the more developed regions contributed 47% of the world's ageing population, compared to the less developed regions, which contributed 53%. Over the span of 50 years those people over 60 years of age decreased in more developed regions and comprised 38% of the world's population, and less developed regions comprised 62%. By the year 2050 projections indicate that more developed regions and less developed regions will constitute 23% and 77%, respectively.

Thus the gap between less developed regions and more developed regions in the ageing population is expected to widen over the decades, with over two-thirds of the world's ageing population originating from less developed regions. According to United Nations (2001), the ageing population in less developed regions will quadruple from 2000 to 2050. This raises important considerations for the African continent given that most countries fall within the less and least developed categories (United Nations 2009).

A comparison of the age structures of the working-age group and the older age group, according to region (Fig. 2.2) shows that, in 2010, youth made up 75% of the world's population, and the elderly 25%. More developed regions in 2010 had a youth population of 45% and less developed regions had a higher

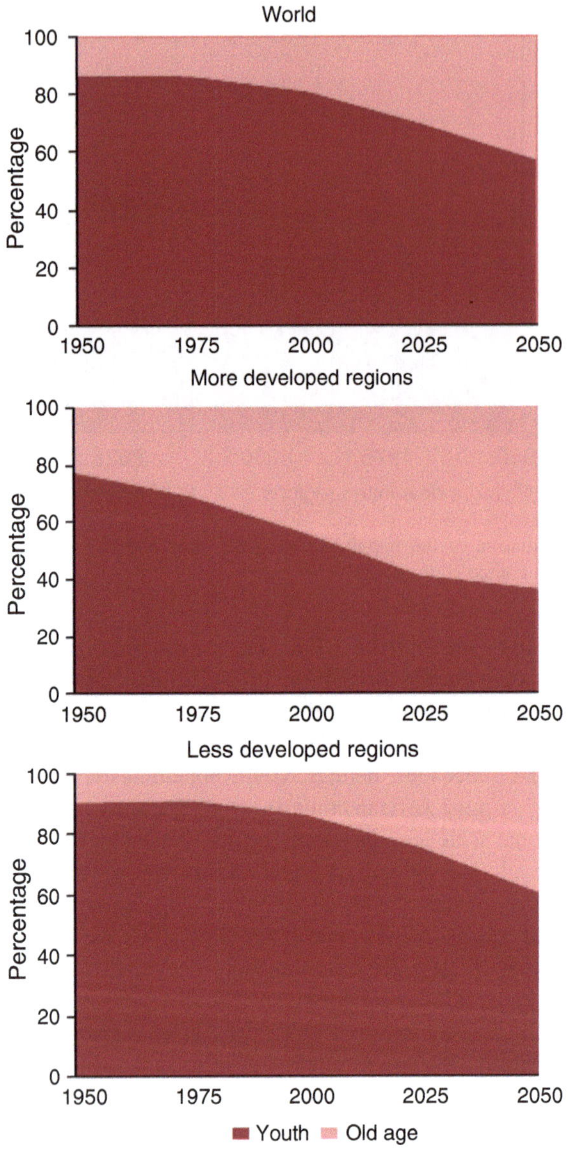

Fig. 2.2 The composition of the youth versus the ageing population by development region, 1950–2050. *Source*: United Nations (2001)

composition of 78%. It is expected that the less developed regions would have greater numbers of youth per elderly person, as the age structure of these nations is younger, especially in Africa.

Projections made for the year 2050 show that the global percentage of youth per elderly person will decrease to 56%, and a decrease is also expected for more developed regions, as the number will decrease to 33%. Less developed countries will also see a decrease; however, the percentage will remain considerably high, with a youthful population comprising 60% of the total population. Thus, whilst the number of ageing persons will be expected to grow over the decades, those of the working age will still proportionately outweigh them in less developed regions.

A disadvantage for the ageing population, as a result of the relatively larger proportion of working-age persons, is the slanting of policies and efforts towards the younger population. However, some argue that the current dependency ratio can work in favour of the elderly if African governments take advantage of the opportunity to maximise employment and economic growth for the working-age population as a means to support the growing number of elderly persons that will be seen in the future.

2.3 Characteristics of the Ageing Population in Africa

It is inevitable that the ageing population of Africa will continue to grow for a number of decades into the future. According to the UNFPA (2008a), the increasing number of ageing people in Africa is a demographic phenomenon linked to decreases in fertility and mortality and is not merely the consequence of economic development. Projections suggest that the number of people over the age of 60 will increase from 64.5 million in 2015 to 103 million in 2030 to 205 million in 2050 (United Nations 2007). These figures indicate that the growth rate of the elderly population will be exponential.

In terms of regional differences across Africa, the fastest rate of growth of the ageing population will occur in Northern and Southern Africa. By 2050, the elderly will comprise 20% of the total population of Northern Africa (Apt 2000). The rate of growth will be much slower in other regions. Regional differences aside, projections suggest that the African population will be growing at a tremendous exponential rate. This is all the more reason to examine the past, present, and future profile of the African ageing population as it will help illuminate the ways that policies and actions can be improved to better suit this demographic phenomenon.

2.3.1 The Demographic Determinants of Population Ageing

Before examining the characteristics of population ageing in the continent of Africa, it is important to understand the demographic determinants of population ageing. The age structure of a population is dependent on the interplay of three main factors: fertility, mortality, and migration rates. However, according to Lesthaeghe (2000), fertility and mortality are far more important demographic factors contributing to the increase in the ageing population than migration. Although the migratory movements of people in and out of the population impacts on the age structure, it is the demographic phenomenon of decreasing fertility and mortality that accounts for the largest growth in the ageing population. An examination of the

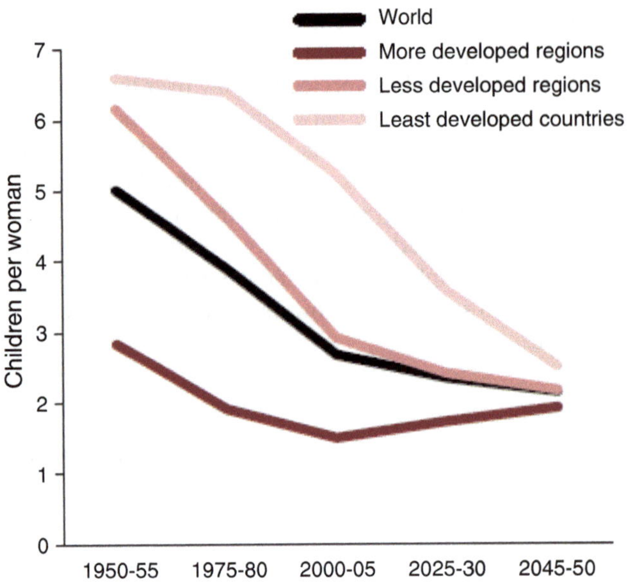

Fig. 2.3 Total fertility rate of the world and development regions, 1950–2050. *Source*: United Nations (2001)

total fertility rate shows a projected decline over a number of decades (Fig. 2.3). Worldwide, the total fertility rate is projected to decline from 5 children per women in 1950–1955 to 2.6 children per women in 2045–2050. In less and least developed countries, the total fertility rate is projected to decrease from 6.6 and 6.4 children in 1950–1955 to 2.6 and 2.3 in 2045–2050, respectively.

The effect of the decreasing fertility rate for the African region, coupled with decreasing mortality, will create an older population in the future. With regard to most developed regions, the total fertility rate is projected to increase from 2001–2005 to 2045–2050. Thus, less and least developed regions are characterised by decreasing total fertility—in contrast to other world regions—which contributes to their increasing ageing population.

The life expectancy of all the world's regions has been increasing over the decades and will continue to do so in the future. A greater life expectancy in a population indicates that the population has control of its mortality rates, and the effect of this will be a larger number of people who survive to older ages. According to Fig. 2.4, the global life expectancy is projected to increase from 45 years in 1950–1955 to 72 years in 2045–2050. More developed regions have always had a higher life expectancy than other world regions and this trend is likely to continue. Less and least developed regions are also projected to increase their life expectancy. In contrast to more developed regions, less and least developed regions have a far lower life expectancy. However, life expectancy is expected to increase for all the major regions and will contribute to a greater proportion of elderly people worldwide in future decades.

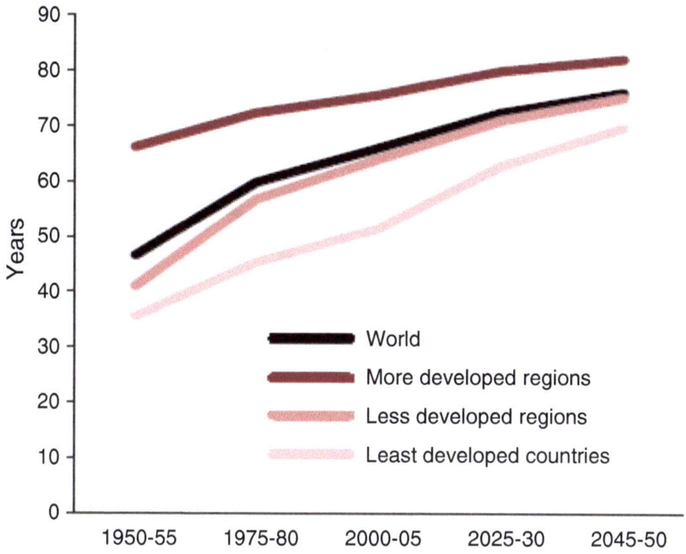

Fig. 2.4 Life expectancy at birth in the world and development regions, 1950–2050. *Source*: United Nations (2001)

Future trends indicate that more developed regions, as well as less and least developed regions, will all continue to increase their life expectancies. Despite this, the range between the different regions will decrease. Projections for 2040–2050 indicate that the global life expectancy will reach 72 years, more developed regions will increase to 83 years, and less and least developed regions will increase to 72 and 65 years, respectively. All this confirms that in the future there will be higher numbers of elderly people in the world.

2.3.2 *The Magnitude and Pace of Ageing in Africa*

Africa consists mainly of the less and least developed countries of the world, and according to the Human Development Report by the United Nations (2009), almost all the nations ranked in the bottom 25 are situated on the continent. African countries are experiencing an increase in the ageing population at a much faster growth rate than more developed regions. During the period 1950–1955, the annual growth rate of the population aged 60 years and over in more and less developed countries was almost the same, but over the next period, 1975–1980, the growth rates in less and least developed countries continued to increase, but decreased in more developed countries (Fig. 2.5). Currently the average annual growth rate of the population aged 60 years and over is 2.8% in less developed countries, 3% in least developed countries, and 0.8% in more developed countries regions. In fact, the less and least developed countries have an annual growth rate which is more

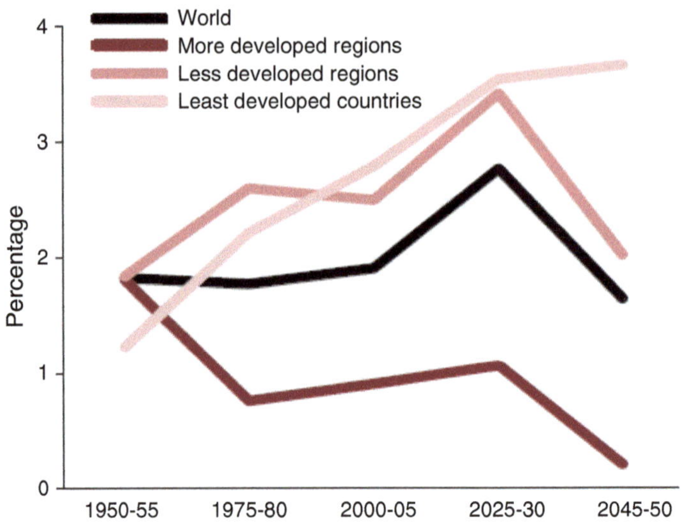

Fig. 2.5 The average annual growth rate of the population aged 60 years+, 1950–2050. *Source*: United Nations (2001)

than 3 times that of more developed countries. This highlights how much faster the annual growth rate of the ageing population of African countries are in comparison with more developed regions. By 2045–2050, less developed regions will have an annual ageing growth rate of 2%, compared with 0.2% in more developed regions. This indicates that less developed regions will have an annual ageing growth rate that is ten times as high as that in the more developed regions. For the least developing regions their annual ageing growth rate will continue to rise in the future, and by 2045–2050 will have the highest rate at 3.7%. For Africa, this means that the increase in the growth rate of the ageing population is a demographic inevitability in the coming decades.

There are striking variations in the age structure of the populations in the more developed regions and the less developed regions. According to Fig. 2.6 from the United Nations (2004), the percentage of individuals aged 60 and over has increased over the decades. In 1950, 9% of the African population was 60 years and over and this is projected to increase to 20% by the year 2020. Projections indicate that by the year 2050, 24% of all Africans will be 60 years and over. In other words almost a quarter of the African population is projected to be elderly. Again this highlights the exponential growth of the ageing population and the changes it will make to the age structure of the African continent. For the African continent, the increase in the ageing population, according to the UNFPA (2008a, b), is due to the large proportion of the working-age population who will in the future contribute to the ageing population. As of 2008, the working-age population in less developed regions stood at 2.5 billion and is expected to increase over time reaching 3.6 billion by 1950 (UNFPA 2008a, b).

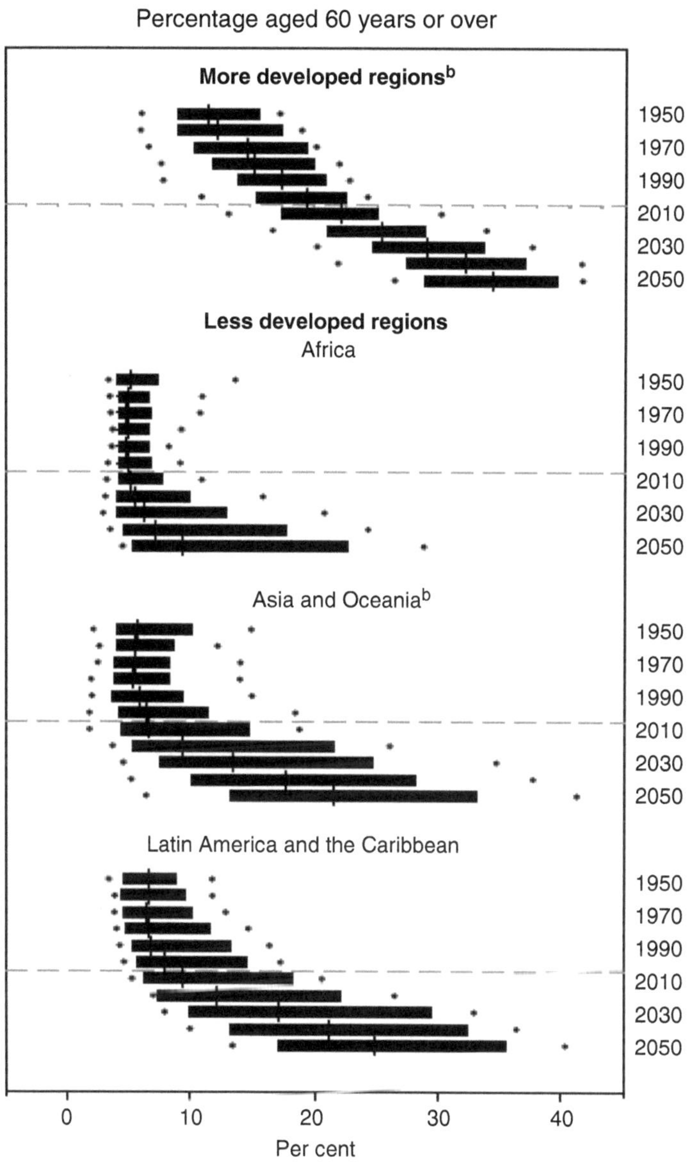

Fig. 2.6 Proportion of the population aged 60 years+, 1950–2050. *Source*: United Nations (2004)

In contrast to this, in the more developed regions, which are mostly made of European countries, a steady increase in the percentage of the population aged 60 years and over will be seen. The percentage of the ageing population is projected to increase according to Fig. 2.6, but this is not a result of population growth. It is clear from Fig. 2.5 the annual growth rate of the ageing population is declining in

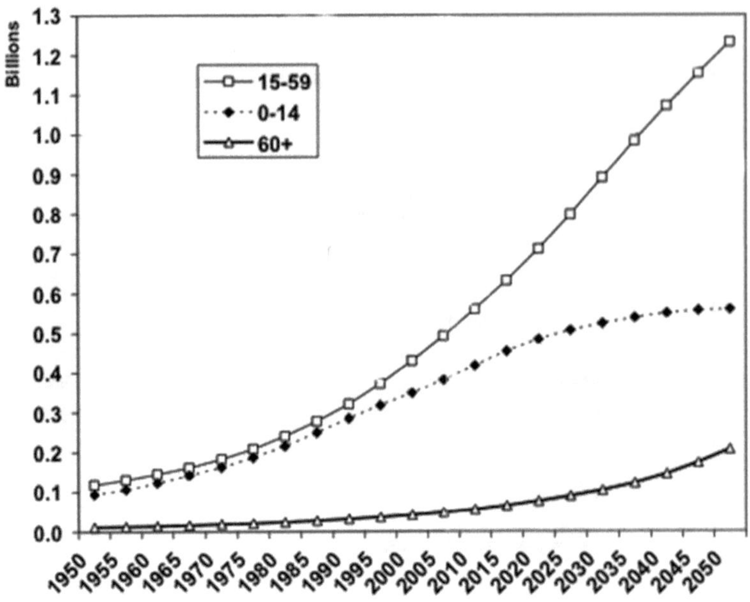

Fig. 2.7 Growth of the African population by age groups. *Source*: United Nations (2007)

more developed regions. According to the UNFPA (2008a, b) the increased proportion of the elderly population in more developed regions will not be due to the population growth of its ageing population, but rather other population phenomena such as migration and improvements in life expectancy associated with high levels of development.

Figure 2.6 shows the distribution of countries according to the percentages of the population aged 60 and over by development region. While more developed regions will experience a further growth in the ageing population, this will not present a major problem since most of the countries in this region are at an advanced stage of the demographic transition, and therefore have greater control of their fertility and mortality rates. According to the UNFPA (2008a, b), over time, the ageing population in more developed regions will stagnate and is not expected in the future to contribute substantially to global population ageing. However, for less developed regions, including most of Africa, the growing ageing population will continue to rise over the next few decades, largely due to the large proportion of the population in the working ages. This demographic phenomenon has the added burden of decreasing fertility, which is foreseen for Africa, from 2.73 children per woman in 2005–2010 to 2.05 in 2045–2050 (UNFPA 2008a, b). The reduction in fertility is beneficial to a nation's development, but will cause population ageing to increase.

Figure 2.7 illustrates that the age structure of the African population has been changing over time. The absolute size of three age categories are shown; these are the dependent groups aged 0–14 years and 60 years and above, and the working-age group of 15–59 years. The working-age group in Africa had a population size of around 0.13 billion in the year 1950 and over time increased to approximately 0.48 billion people. It is projected that by 2050 this group will comprise the majority of the African population, almost 1.23 billion. However, this is not surprising, since it is widely recognised that the African population is youthful.

Although the working-age population will form the majority of the African population, it is important to note that the ageing population will also continue to steadily grow in numbers. In 1950, the African population aged over 60 years was around 0.03 billion and is projected to increase to 0.14 billion by 2050. This may not appear to be significant growth in population size when analysed in billions, but over 100 years from 1950 to 2050 the ageing population will have multiplied almost five times. This significance can be highlighted by comparing Africa to Europe. The ageing population of Europe is projected to grow from 10% in 1950 to 33% in 2050 (European Centre for Social Welfare Policy and Research 2008). Over the period of 100 years, from 1950 to 2050, the older population in Europe is expected to increase threefold.

2.4 The Profile of the African Ageing Population

Projections reveal that the number of people aged 60 years and over is expected to rise over the next few decades. Table 2.1 indicates that the number of people aged 60 and over was 53,770,000 in 2009, and this is projected to increase to 212,763,000 in 2050 (United Nations 2008a). In 2009, the population aged 60 and over accounted for 5% of the total African population, and this is expected to increase to 11% by 2050. Furthermore, the oldest age group, those 80 years and older, constitute 8% of the African population and will account for 10% of the population in 2050.

An important variable to consider when analysing data on the ageing population is gender. Studies conducted by various organisations including the World Bank stress the importance of gathering data on gender differentials, as the continuation of gender-based inequality serves as a hindrance to Africa's growth (Gelb 2001). Table 2.2 shows that the sex ratio of the African population is such that there are 85 men per 100 women.

Thus, in Africa there are more women aged 60 years and over than men. This also applies to the oldest age groups who are 80 years and above, with 68 men for every 100 women, or in other words, slightly more than 25% females than males. This is expected, given that most studies suggest that women live longer than men,

Table 2.1 The African population aged 60 years and over, 2009 and 2050

	2009	2050
Number (thousands)	53,770	212,763
Percentage of total population	5	11
Share of persons 80 years or over	8	10

Source: United Nations (2008a)

Table 2.2 The African population sex ratio, 2009

Sex ratio (men per 100 women)	
60 years+	80 years+
85	68

Source: United Nations (2008b)

Table 2.3 Gender differentials of the African population

	Men	Women
Life expectancy at age 60, 2005–2010	15	17
% Married[a], 60 years+	85	39
% Living alone, 60 years+	6	11
% Participating in labour force, 60 years+	61	34

[a]"Married" includes those in consensual unions
Source: United Nations (2008b, c)

and thus they form the majority in the elderly category. It is important to ascertain whether policies are attuned to the predominantly ageing female population.

Further examination of gender differentials reveals that the life expectancy of women aged 60 years is higher than men (Table 2.3). Women aged 60 are expected to live an additional 17 years, and men an additional 15 years. With regard to marital status, a far greater percentage of elderly men are married than women. Almost 85% of men aged 60 years and over are either married or in a consensual union, compared with 39% of women. Thus, elderly men are more likely to be married or in a consensual union than women.

According to Waite (1995), the benefit of analysing marriage statistics is that it gives a good indication of the socio-economic status of households, and since married individuals have better health than non-married individuals, it serves as an indicator of a population's health status. Considering the large discrepancy between the number of married elderly males and the number of married elderly females, it appears that gender inequalities are highly prevalent in African society.

Elderly men are in an advantaged position, since most of them are married, and by being married they enjoy better health and a better socio-economic position than the 15% of men who are not. As for women, the majority are disadvantaged in terms of marital status, as only 39% are married or in a consensual union. This means that 61% of elderly African females have a decreased health and socio-economic status in relation to the remaining elderly females. In addition, there are more elderly women who live alone than elderly men. There are 6% of elderly men living alone compared with 11% of elderly women. With regard to economic activity, men aged 60 and over are also more economically active than their female counterparts. In Africa, among those aged 60 years and over, 61% are males. This percentage is much lower for females, with only 34% of women aged 60 years or older still economically active.

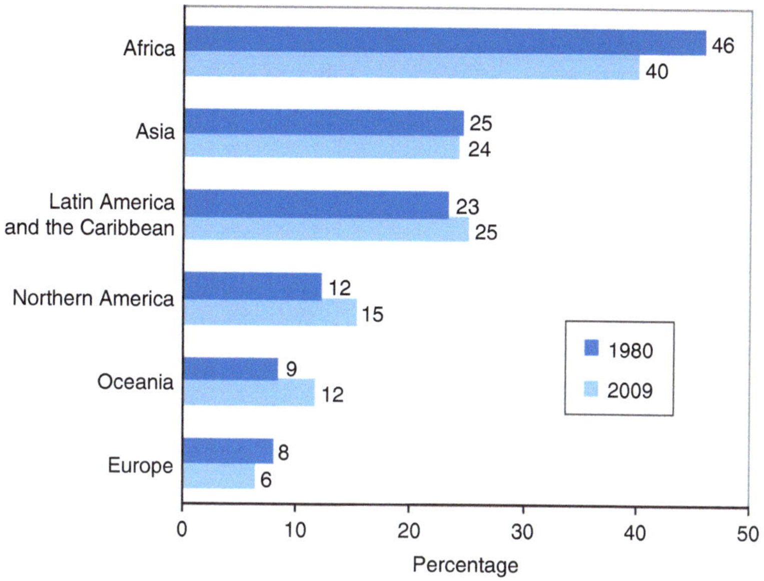

Fig. 2.8 Labour force participation of those 60 years and over, 1980 and 2009. *Source*: United Nations (2009)

Figure 2.8 shows that Africa remains the most economically active ageing population. In 1980, 46% of the population aged 60 years and above were participating in the labour force, but this declined to 40% in 2009. However, labour force participation by the older population still remains higher in Africa than other regions of the world. What this implies is that the elderly in Africa are continuing to participate in the labour force partly due to difficulties in securing a pension in their old age.

According to Lam et al. (2006), employment rates are higher amongst the elderly who do not receive a pension, and considering that Africa has the most economically active elderly population, this suggests that they are more likely to participate in the labour force in the absence of pension benefits.

When the ageing population of Africa is disaggregated according to place of residence, it is clear that there is only a minor difference in elderly people living in rural and urban areas (Table 2.4). According to UNDP (2005), 6% of the population who live in rural areas in Africa are aged 60 years and over, while 5% of the population who live in urban areas in Africa are aged 60 years and over. Thus, the ageing population is almost equally distributed in urban and rural areas. One would expect a far greater proportion of the elderly to be concentrated in the rural areas, since Africa is known to be a developing continent. Asia also has an equal distribution of elderly people in urban and rural areas. Northern America, like Africa, has a slightly greater percentage of elderly who live in rural areas. However, Europe has the highest proportional difference, with far more elderly people found in rural areas than in urban areas, 21% and 17%, respectively.

Table 2.4 Percentage of the population in urban and rural areas aged 60 years and over

	Urban	Rural
Africa	5	6
Asia	9	9
Northern America	16	18
Europe	17	21

Source: UNDP (2005)

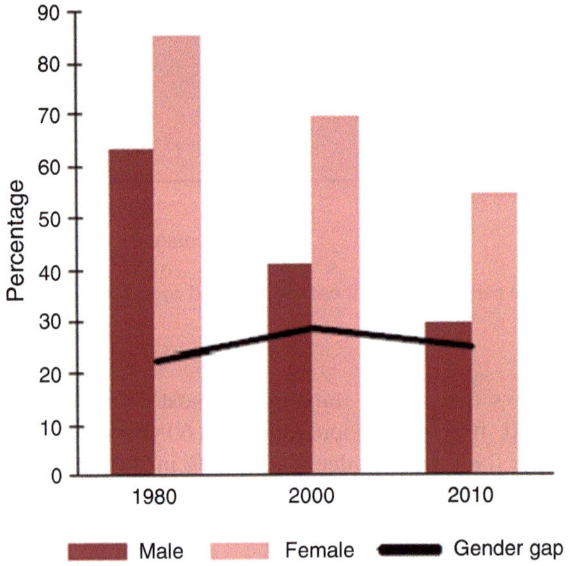

Fig. 2.9 The male and female illiteracy rate for those aged 60 and over and the gender gap, in less developed countries, 1980–2010. *Source*: United Nations (2001)

Earlier projections of the level of education of elderly Africans (Fig. 2.9) suggest that by 2010 less than half of all Africans aged 60 years and older will be illiterate. Although the illiteracy rate is expected to decrease over the decades, there is likely to remain a substantial number of illiterate people in this age group. Furthermore, levels of illiteracy are higher among women in this age group than men. Again, women are in a disadvantageous position on the basis of their gender. The total percentage of elderly Africans who are illiterate is 43% (57% literate), in addition, as discussed previously almost half of all elderly Africans participate in the labour force (Fig. 2.8). The relationship between literacy and employment confirms the findings of Lam et al. (2006) that literacy and employment have a positive correlation and an almost equal percentile.

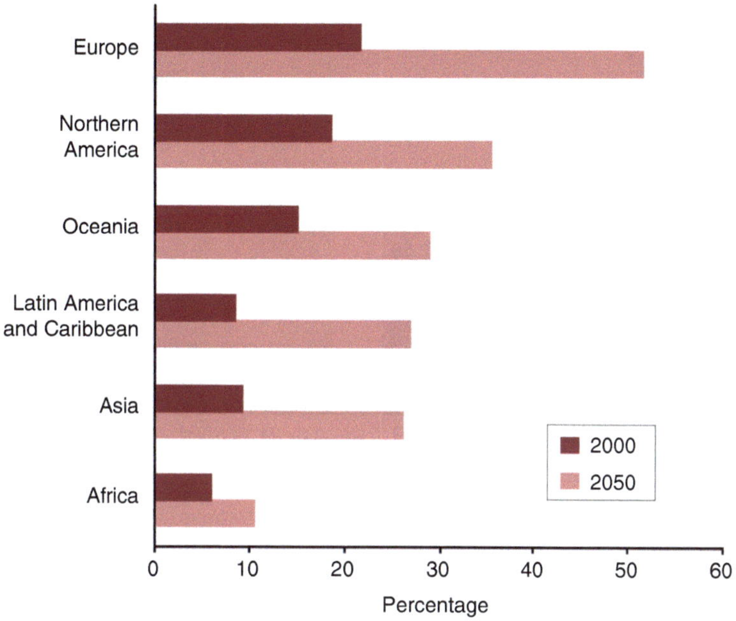

Fig. 2.10 Elderly dependency ratio by area, 2000 and 2050. *Source*: United Nations (2001)

2.4.1 The Dependency of the Ageing Population on the Working-Age Population

One would expect to find that the number of people of working age in Africa outnumbers the ageing population, since Africa is known to have a youthful population. According to data from the United Nations (2008a), the age pyramid of the African population is comprised mostly of the working-age population. Furthermore, it is important to ascertain the proportional difference or ratio of the working-age population to the ageing population. These figures are important since it will help determine how dependent the ageing population is on the working-age population for their social and economic needs. The dependency ratio shown in Fig. 2.10 indicates that in 2000, 5% of the elderly in Africa were dependent on the working-age population (United Nations 2001). In contrast, in 2000, 20% of the elderly in Europe were dependent on the working-age population. Figure 2.10 shows that in Africa the elderly dependency ratio is on the increase. However, the other continents have a higher dependency ratio than Africa. This paints a relatively favourable situation for the African ageing population, as the elderly dependency ratio is much smaller than the wealthier continents.

The projections show that support for the ageing population will decline in the future. The percentage of elderly to the economically active population will increase over time and will reach 10%. The figure will double over the course of 5 decades.

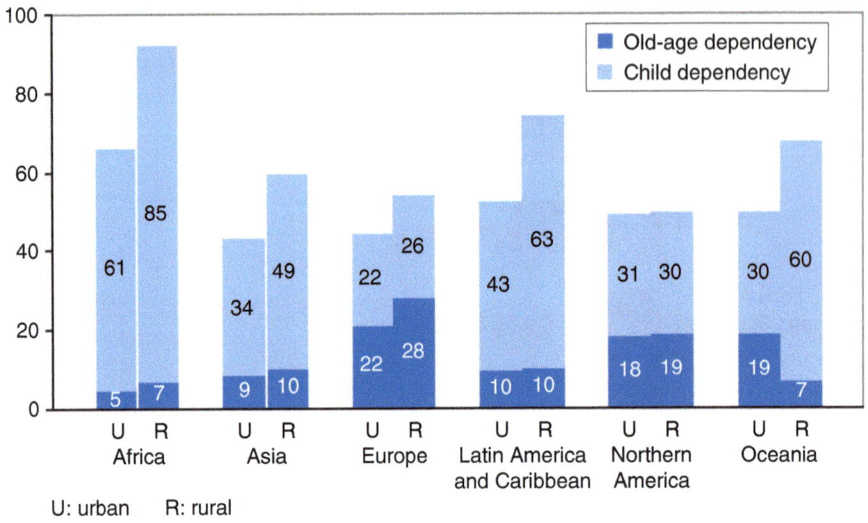

Fig. 2.11 Old age dependency percentage versus child dependency percentage in urban and rural areas. *Source*: United Nations (2009)

The increase in the dependency ratio will have negative implications for the African ageing population, especially if the increase continues in the years beyond 2050. The ageing population will not receive as much support in the future from the working-age population because they will also increase in numbers, which will change the age structure, and thus increase the dependency ratio.

Recent figures by United Nations (2009), Fig. 2.11, indicate that the ageing population of Africa is far more dependent on the working-age population than children aged 0–14 years. The elderly who live in rural areas comprise 85% of all those who are dependent, compared with children who constitute only 7%. In urban areas, the elderly comprise 61% of all those who are dependent, compared with children who constitute only 5%.

Unlike Africa, in Europe the elderly and children are both equally dependent on the working-age population, which implies that they place a similar "burden of responsibility" on the working-age population. The elderly in Africa, on the other hand, place a far greater burden of responsibility than children on the working-age population. The elderly, who form the proportional majority, are not extensively acknowledged on the developmental front, as attention is often paid to children; this results in the vast majority of elderly in Africa being excluded from national priorities.

2.4.2 Regional Trends in Ageing

The following section examines the trends in ageing in the African continent. Population ageing in Africa varies substantially by region. The regions will be classified under five main categories: Northern Africa, Southern Africa, Eastern

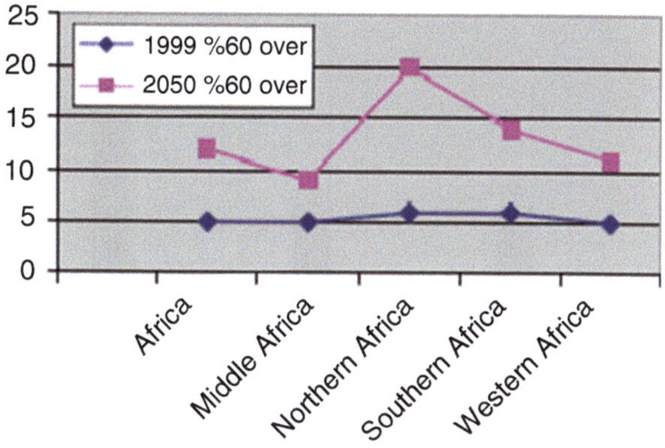

Fig. 2.12 Increase in the percentage of the ageing population in Africa, by region. *Source*: Kalasa (2004)

Africa, Western Africa, and Central/Middle Africa. Each region is characterised by the trends of the countries within it, and thus some regions differ substantially from others based on which countries fall into its category. Specific country-level trends will be discussed in the next section, but before zoning in on countries, it is important to examine the five African regions, in order to ascertain the differences and similarities in ageing trends.

Projections indicate that, of all the regions, the largest increase in the ageing population will occur in Northern Africa. According to Fig. 2.12, in 1999, 6% of the Northern African population was 60 years and over, and this is expected to increase to 20% in 2050. This is almost a fifth of the Northern African population and will represent the greatest proportion of ageing adults within all the five African regions.

Southern Africa will also witness a steady increase in its ageing population, from 6% in 1999 to 14% in 2050. However, the sharpest increase in the ageing population will occur in Northern Africa. The ageing population in Southern Africa is expected to double over the next 5 decades, whereas the ageing population of Northern Africa is expected to treble over the next 5 decades. The ageing population of Western Africa will also increase from 5% in 1999 to 11% in 2050, almost doubling over this period. The smallest increase will occur in Middle (Central) Africa, from 5% in 1999 to 8% in 2050.

When the regional comparison is shifted to consider the sub-Saharan region in relation to gender, it can be seen that males in the region fare far better than females. Figure 2.13 illustrates the gender differences by place of residence and living arrangements in the sub-Saharan region. Only 6% of elderly men in rural areas, and 7% of elderly men in urban areas live alone, compared to 13% of elderly women in rural areas and 10% in urban areas. Elderly women thus face a greater risk of loneliness and abandonment than their male counterparts at older ages.

The majority of elderly men live with at least one child, 69% in rural areas and 73% in urban areas. Fewer elderly women live with at least one child, 50% in rural areas and 55% in urban areas. Elderly women are also tasked with more familial and household responsibility than elderly men, as the percentage having to fend for their

Fig. 2.13 Living arrangements of persons aged 60 and over in sub-Saharan Africa around 2003 by place of residence. *Source*: Kinsella and Wan (2009)

grandchildren, and the percentage without assistance from the grandchild's parent is both higher for women.

From these figures one can assume that the AIDS epidemic in sub-Saharan Africa is impacting the living arrangements of the elderly. At a demographic level, the younger cohort has a high mortality rate due to AIDS, resulting in a higher proportion of elderly people compared with the younger population (Lloyd-Sherlock 2000). At a social and household level, often when the child of the elderly person has died or is severely ill due to HIV/AIDS, the elderly person is tasked with the responsibility of caring for, or even raising the grandchild. However, elderly women are more disadvantaged by this situation than elderly males, as they have a higher percentage of skipped generation households. Thus elderly women have to assume far greater responsibility than elderly men and they are also likely to be alone and have less support from those around them than their male counterparts.

At a regional level (Fig. 2.14), the highest concentration of elderly men is in Eastern, Northern, and Western Africa, with 46%. This figure is similar to other regions in the world such as South-eastern Asia, Western Asia, the Caribbean, Central America, and South America. The percentage of the population that is men aged 60 and over is 44% in Middle Africa and 40% in Southern Africa. However whilst the gender composition of most African regions is similar to other world regions, the impact of the gender imbalance at older ages in Africa can pose a problem for females who are often alone in their old age, since Africa is home to twenty of the world's worst health systems (Robinson 2007). Thus, inadequate health systems in Africa will negatively impact females in their old age to a much larger extent than in other world regions. It is important to take this gender imbalance into consideration when designing ageing policies.

At the older ages, there are a high percentage of women than men. In Africa, women as a group tend to live longer than men, and this means that they are likely

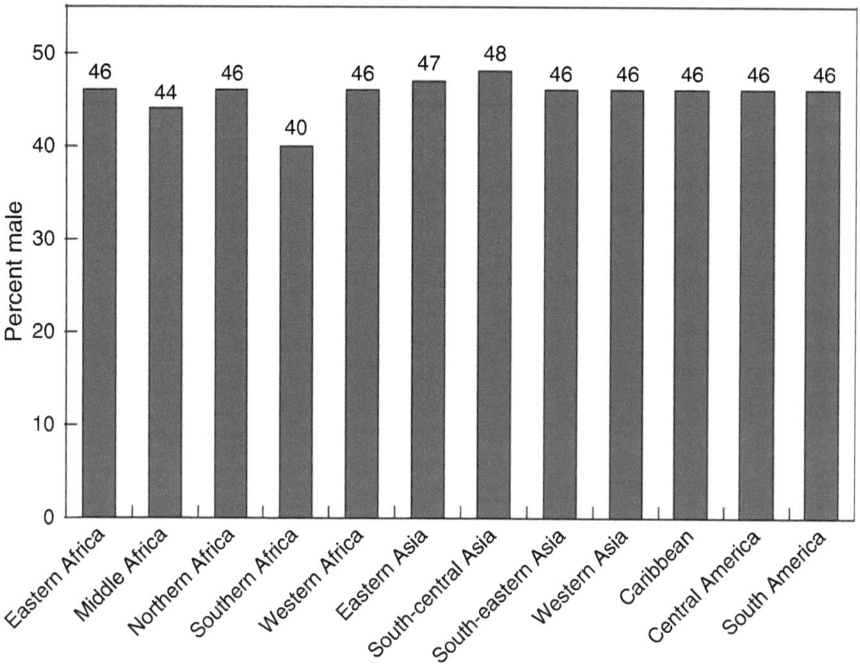

Fig. 2.14 The percentage of males amongst the population aged 60 years and over by region. *Source*: Knodel and Ofstedal (2003)

to outlive their male partners, and, as a result, spend more of their time living alone. Knodel and Ofstedal (2003) note that the extent of difficulties facing elderly women is heavily dependent on their social and economic context; however, they note that their widowhood negatively impacts on their lives nevertheless.

Table 2.5 shows that the life expectancy in Africa is much lower than other continents for 2000–2005 at 51.3 years, but it is projected to increase to 69.5 years for 2045–2050. This indicates that Africans will have a longer life expectancy in the future, thus the increase in the ageing population will become a trend of the continent as a whole. In particular, men and women aged 60 and 65 years will experience the highest increase in life expectancy of all the elderly in Africa. Both elderly men and women are expected to see a large increase in their life expectancy over the next few decades. The life expectancy of men and women will rise from 50.5 years and 52.1 years in 2000–2005 to 67.9 years and 71.2 years in 2045–2050, respectively.

Over the next few decades the largest increase in the life expectancy of all the African regions will occur in East Africa, from 45.4 years in 2000–2005 to 67.2 years in 2045–2050. The growth in life expectancy will occur in other regions, but it will be a more gradual process. With regard to gender differentials, females have a higher growth in life expectancy than males in all African regions. This indicates that females will continue to outlive males into old age, and if gender inequalities persist into the future, this will only magnify the problems that elderly women face in

Table 2.5 Life expectancy by sex and particular ages in Africa and its regions

	Total				Female				Male			
Region	Birth	60	65	80	Birth	60	65	80	Birth	60	65	80
Africa												
2000–2005	51.3	16.2	13.1	5.8	52.1	17.0	13.7	6.1	50.5	13.3	12.4	5.5
2045–2050	69.5	20.2	16.4	7.3	71.2	21.4	17.4	7.9	67.9	18.9	15.3	6.6
East Africa												
2000–2005	45.4	15.6	12.7	5.7	46.0	16.3	13.3	5.9	44.8	14.8	12.2	5.4
2045–2050	67.2	19.7	16.0	7.1	68.7	20.6	16.7	7.4	65.8	18.7	15.2	6.7
Middle Africa												
2000–2005	50.0	16.0	12.9	5.7	51.1	16.7	13.4	5.9	48.8	15.3	12.4	5.4
2045–2050	69.0	19.7	16.0	7.0	70.4	20.5	16.6	7.3	67.6	18.9	15.3	6.7
Northern Africa												
2000–2005	66.4	17.1	13.6	5.6	68.0	17.9	14.3	5.9	64.8	16.1	12.8	5.3
2045–2050	77.0	21.0	17.0	7.5	79.3	22.7	18.5	8.4	74.7	19.2	15.3	6.5
Southern Africa												
2000–2005	46.4	15.6	13.0	6.3	47.1	17.6	14.6	6.8	45.6	13.2	11.0	5.3
2045–2050	66.2	20.6	17.1	8.4	67.2	23.1	19.2	9.3	65.1	18.1	14.7	6.8
Western Africa												
2000–2005	51.3	16.3	13.2	6.1	51.8	16.9	13.6	6.3	50.7	15.7	12.7	5.8
2045–2050	69.0	19.9	16.1	7.1	70.4	20.7	16.8	7.4	67.7	19.0	15.4	6.8

Source: United Nations (2007)

their old age. Knodel and Ofstedal (2003) argue that, instead of focusing on the challenges facing elderly women, more attention should be directed at improving male survivorship as a means to reduce the number of elderly widows. As stated by Van Dulleman (2006), old age is a gendered experience; however, it is important to consider the elderly male perspective as it may serve as an effective conceptual tool to help reduce gender inequalities and the challenges facing the elderly. Taking the elderly male perspective into consideration would prove helpful during policy-making and at a programmatic and developmental level, as the continual focus on elderly women's issues emphasises a "victimised" ideology which may in fact perpetuate their issues.

2.5 Demographic Trends in Ageing

At a regional level, Fig. 2.15 shows that Northern, Eastern, and Western Africa have similar numbers of elderly men and women, at 14,758,000, 14,996,000, and 14,332,000 aged 60 years and over in the year 2009, respectively.

As of 2009, the region with the smallest ageing population was firstly Southern Africa with 4,013,000, followed by Central Africa with 5,671,000. By 2050, Northern Africa will have the highest concentration of elderly people in the continent. Projections suggest that the ageing population of Northern Africa will

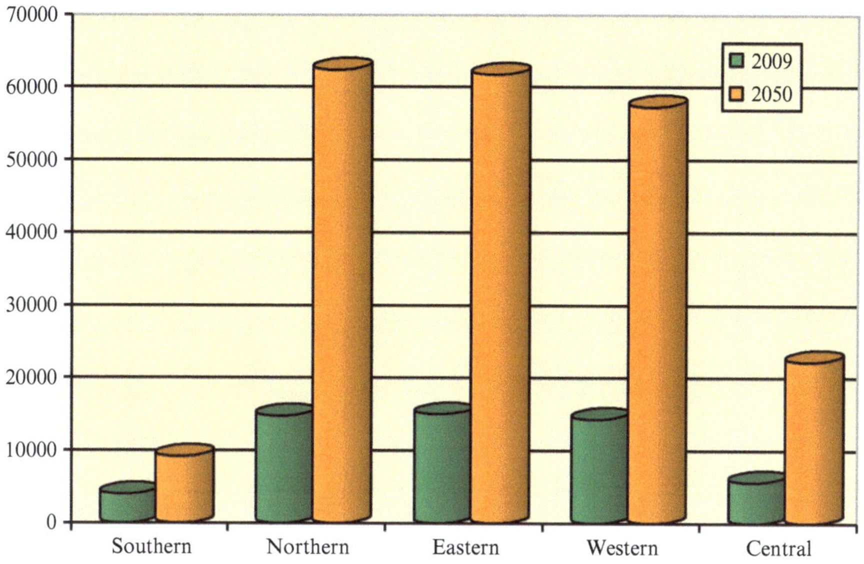

Fig. 2.15 The ageing population in thousands, 60 years and over. *Source*: United Nations (2008a)

increase almost fourfold to 62,388,000 in 2050. This will be followed by Eastern Africa which will increase to 61,740,000 and Western Africa to 57,251,000 in 2050.

While not as large an increase as the other regions, the population aged 60 and over in Central Africa will increase from 5,671,000 in 2009 to 22,181,000 in 2050, and the population aged 60 and over in Southern Africa will double from 4,013,000 in 2009 to 9,203,000 in 2050.

The ageing population constitutes 7% of the total population of Southern Africa and Northern Africa, according to 2009 figures (Fig. 2.16). In Eastern, Western, and Central Africa, the elderly constitutes 5% of the total population.

In Northern Africa, the ageing population will constitute the largest share of the total population compared to all African regions. The ageing population will comprise 19% of the total population by the year 2050. By 2050 the ageing population of Southern Africa will comprise 14% of the total population. In addition, by 2050 the ageing population of Eastern Africa and Western Africa will increase to 9% of the total population, and Central Africa to 8% of the total population.

The population aged 80 years and over constitutes 8% of the Southern African population in 2009 and will almost double to 15% in 2050 (Fig. 2.17). The increase in the size of the oldest demographic group concurs with previous observations in this chapter that the life expectancy of the ageing population is increasing in Africa. In Northern Africa the life expectancy will also increase over the decades, as the share of the population aged 80 years and over will rise from 9% in 2009 to 13% in 2050.

In the future, the oldest cohort, the proportion of the population aged 80 and above in Eastern, Western, and Central Africa, is expected to increase only slightly

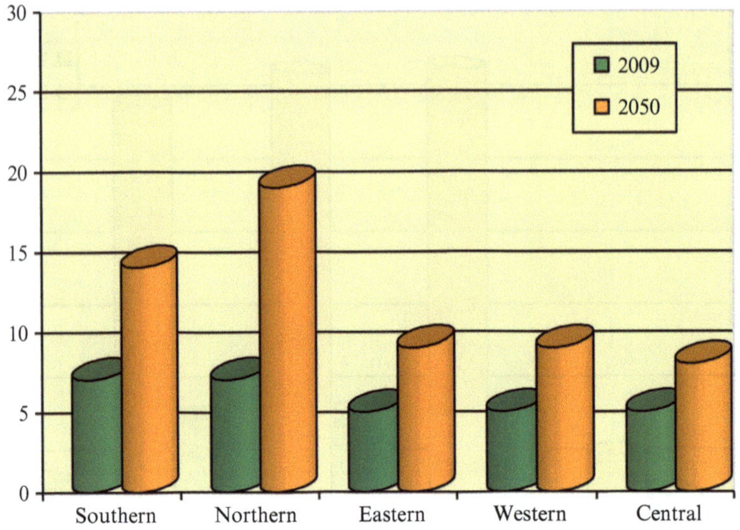

Fig. 2.16 Percentage of the total population, 60 years and over. *Source*: United Nations (2008a)

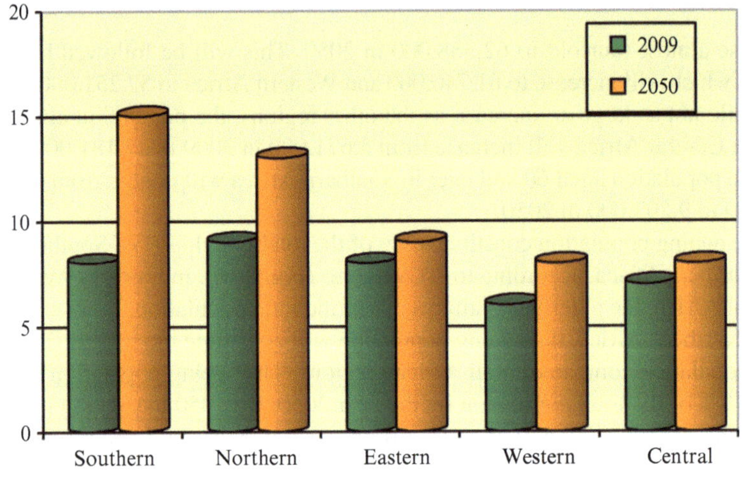

Fig. 2.17 Share of the total population aged 80 years and over. *Source*: United Nations (2008a)

and constitute a slightly larger share of the total population than in 2009. Over the next few decades the life expectancy of the oldest cohort in Eastern Africa will not alter much. In 2009, 8% of the Eastern African population was aged 80 years and over, and by the year 2050 it will increase by just one percent to 9%.

In Western Africa the percentage of the oldest old group will increase from 6% in 2009 to 8% in 2050. Thus, life expectancy of the oldest cohort will not see much change in this region over the decades. Similarly, Central Africa will experience

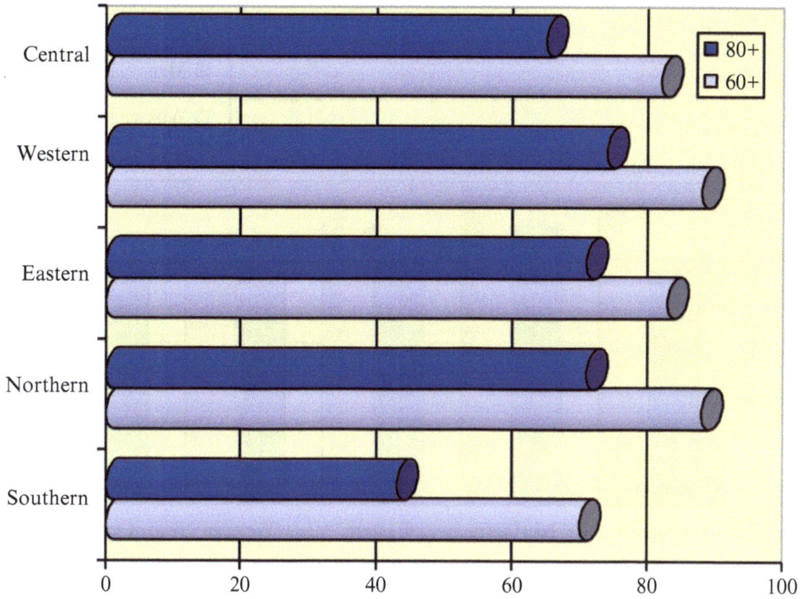

Fig. 2.18 The sex ratio of the ageing population. *Source*: United Nations (2008b)

only a slight increase in the share of population aged 80 years and above, from 7% in 2009 to 8% in 2050.

The sex ratios (Fig. 2.18) indicate that Western Africa has the highest sex ratio for both cohorts 60+ and 80+. This is followed by Northern and Eastern Africa. The figures for Western Africa indicate that there are 88 men for every 100 women aged 60 years and above. This figure decreases slightly for the oldest cohort, with 74 men for every 100 women aged 80 years and above. Similar figures are seen in Northern Africa with 88 men for every 100 women. This figure decreases for the oldest cohort, with 71 males for every 100 women aged 80 years and above.

In Eastern Africa, there are 83 men for every 100 women aged 60 years and above. For the oldest age bracket, those aged 80 years and above, there are 71 men for every 100 women. In Central Africa there are 82 men for every 100 women aged 60 years and above, this ratio decreases slightly with increasing age, with 65 men aged 80 years and over for every 100 women.

Southern Africa has the lowest sex ratio, with far fewer men than women. In 2009, for every 100 women aged 60 years and over, there were 70 men. In other words, there were 30% more women of this age than men. This indicates the difficulties that women of this age are likely to face, since their male partners are less likely to survive as long as they do, and as a result they spend most of their elderly years without a male partner. This issue only worsens with age, as there are only 43 men for every 100 women aged 80 years and over. Thus more than half of all women in the oldest cohort are without a male partner companion in Southern Africa.

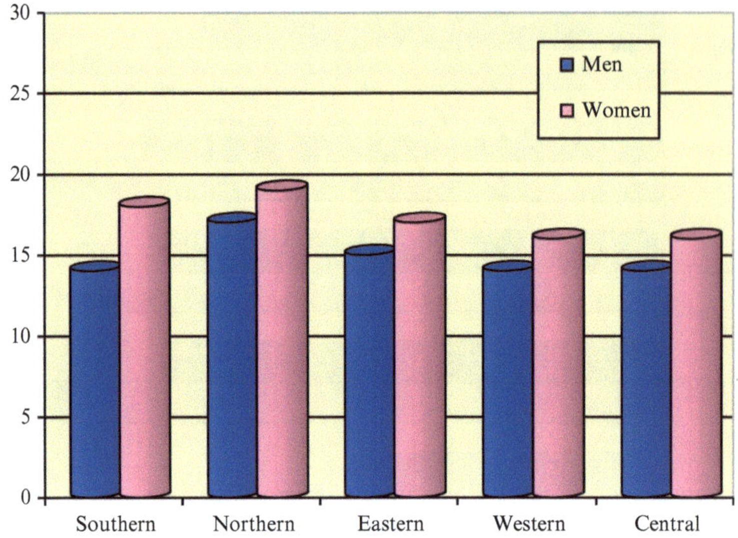

Fig. 2.19 Life expectancy at age 60, 2005–2010. *Source*: United Nations (2008b, c)

2.5.1 Gender Differentials

Figure 2.19 shows the life expectancy for both men and women aged 60 years and over, by region. Of all the African regions, the life expectancy for men and women aged 60 years and over is highest for Northern Africa; however, the regional differences are small. What is common across all regions is that elderly women have a longer life expectancy than men; however, the gender differentials are not large. In Northern Africa, men aged 60 years and over are likely to live an additional 17 years, compared with women with a slightly higher life expectancy of an additional 19 years. In Eastern Africa, men are likely to live a further 15 more years after reaching age 60, whereas females are expected to live a further 17 years.

In Central Africa, the life expectancy at age 60 for women is higher than for men. The life expectancy is 16 years for women and 14 years for men who live to 60 years. Similarly, in Southern Africa the life expectancy for women is higher than that of men. According to Fig. 2.19, men are likely to live an additional 14 years after they turn 60, compared with women who are likely to live an additional 18 years. In the case of Western Africa, the life expectancy of elderly women is slightly greater than men. The life expectancy is a further 16 years for women and 14 years for men who live to 60 years.

With regard to marital status and consensual unions (Fig. 2.20), Northern Africa has the highest percentage of married men aged 60 and above, and Southern Africa has the largest percentage of married women aged 60 and above. In Northern Africa, 90% of all men aged 60 and over are married compared with less than 40% of women. Men aged 60 and over in this region thus have far greater companionship in

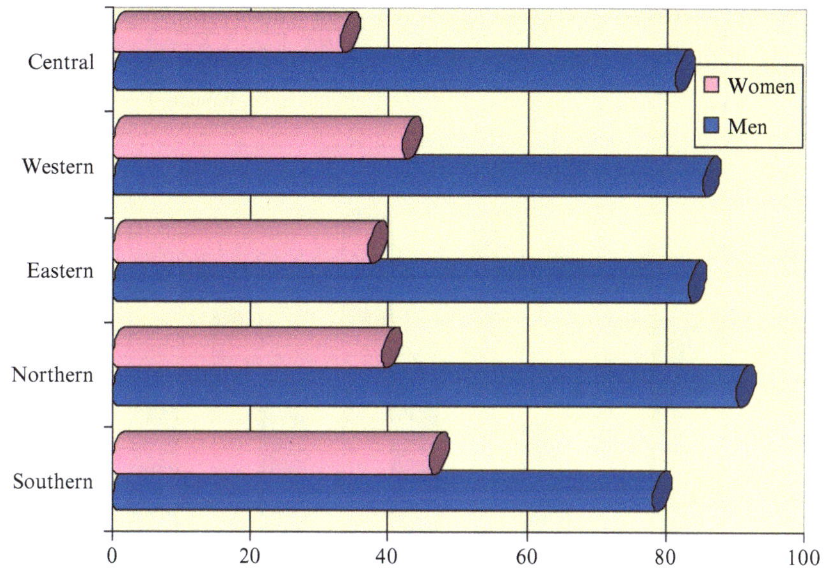

Fig. 2.20 Percentage married, 60 years and over. ("married" includes those in consensual unions). *Source*: United Nations (2008b, c)

their old age than women. Similarly in Western Africa fewer women are married or in a consensual union at 60 years and over. Men are two times more likely than women to be married or in a consensual union at age 60 and over.

In Southern Africa, a far greater percentage of older men (78%) are married or in a consensual union than women (46%). In Eastern Africa, 83% of all older men are either married or in a consensual union, compared with 37% of older women. Similarly, in Central Africa, 81% of older men are either married or in a consensual union, compared with 33% of older women. The figures indicate that, for all African regions, men are more likely than women aged 60 and over to have greater companionship and support as a result of being married in their older age.

Figure 2.21 shows that the percentage of the population aged 60 and over who live alone is highest for women in Central Africa, at 13%, and highest for men in Southern Africa, at 8%. A gender comparison shows that in Southern Africa, despite the percentage of older, married men being much higher than older women, there is an almost equal percentage of men and women living alone. In Southern Africa, 8% of men and women aged 60 and over are living alone. However for Central Africa, around twice as many older females live alone than older males, at 13% and 6%, respectively. Similarly in Eastern Africa twice as many women are alone as men, at 12% and 6%, respectively.

The gender disparity in living arrangements is greatest in Northern Africa. Older men have far greater companionship in their old age compared with older women, and this is corroborated by the percentage living alone. Only 4% of older men live alone, compared with 12% of older women who live alone. In Western Africa almost

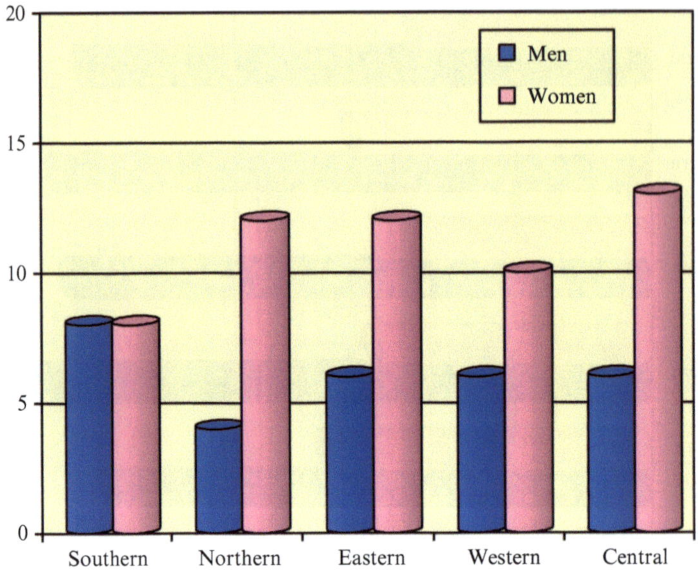

Fig. 2.21 Percentage living alone, 60 years and over. *Source*: United Nations (2008b, c)

twice as many older women (10%) live alone than older men (6%). In all the African regions, apart from Southern Africa, women are more likely to be living alone at older ages than men. At least twice as many older women than older men live by themselves.

In comparison to the other regions, Eastern Africa has the largest percentage of men and women aged 60 years and over in the labour force. Gender comparisons show that, in Eastern Africa, 79% of men, compared with 56% of women, are in the labour force. This is followed by Central Africa where labour force participation is 75% for men and 47% for women aged 60 years and over. The third highest region in terms of labour force participation is Western Africa. Figure 2.22 shows that 65% of elderly men in Western Africa are economically active, compared with only 36% of women.

In Northern Africa, men aged 60 and over are five times more likely to participate in the labour force than women. Only 8% of women aged 60 years and over participate in labour force, compared with 41% of men. The participation of the population aged 60 years and over is lowest in Southern Africa. Men aged 60 years and over are almost two times more active in the labour force than women. Almost 38% of men aged 60 years and over are economically active, compared with 18% of all women aged 60 years and over. In all the African regions, older men are more economically active than older women.

The old-age support ratio is a key indicator of the pressures that population ageing is likely to pose for pension systems. It measures how many people there are of working age (20–64) relative to the number of retirement age (65+). Figure 2.23 shows the old age support ratio by African region.

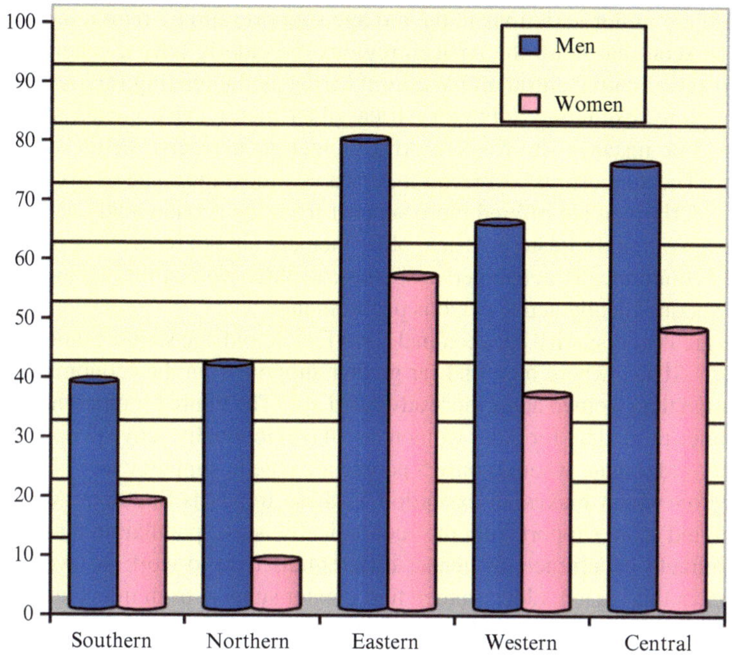

Fig. 2.22 Percentage in the labour force, 60 years and over. *Source*: United Nations (2008b, c)

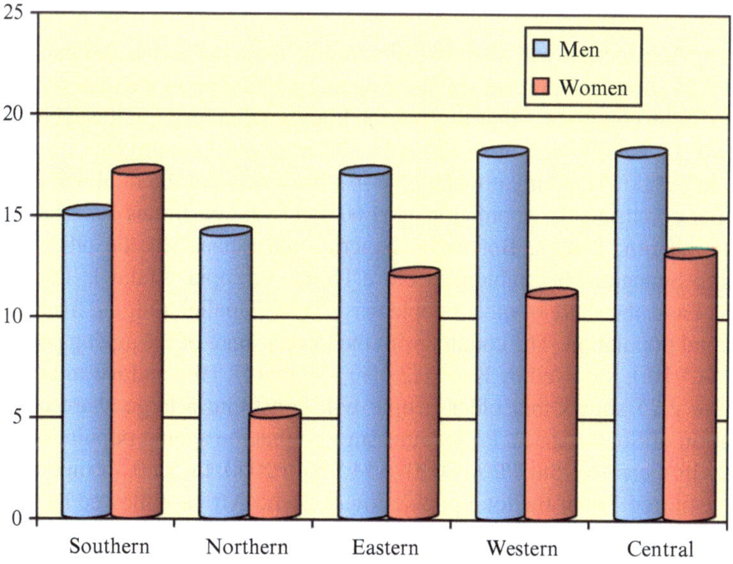

Fig. 2.23 Old age support ratio, 2009. *Source*: United Nations (2008b)

There are minor variations in the old age support ratio by region in Africa, and it can be seen that in all the African regions the elderly have a generally similar support ratio, with Central and Western Africa as the leading regions. However, in all regions, excluding Southern Africa, older men receive greater support than women. For instance, in Western Africa older men receive better support than females. The support ratio of Central Africa is almost the same as Western Africa where elderly men are offered more support from the economically active population than women. In Eastern Africa, older men are able to receive support and care from 17 economically active persons, whereas older women find less support with only 12 economically active persons per woman.

The greatest disparity between male and female old age support ratios is seen in Northern Africa, where men find far greater support from the economically active population than women at the more advanced age. The figure for older men is almost three times more than for older women, as 14 economically active people are found for every older man, whereas only 5 people are able to support every older woman. The region which makes an exception to these trends is Southern Africa where women find more support from the economically active population than men. This is an unlikely occurrence, as gender differentials tend to work in favour of men; however in this region, older women find greater support from the younger population than older men. For each older woman, there are 17 economically active people to provide support, in comparison with older men who have 15 economically active people to support them.

2.6 The Country-Level Differentials in Population Ageing

The following section will analyse differentials in ageing across African countries. Figure 2.24 shows the percentage of the ageing population of the total population in select African countries. Mauritius has the highest percentage of the ageing population with 9.5%, followed by South Africa (7.8%) and Lesotho (6.9%). In Central Africa Republic the ageing population constitutes 6.1% of the total population.

In four countries, the population aged 60 and over constitutes 5% or more of the total population: Eritrea, Botswana, Nigeria, and Ghana. In the other countries, including Senegal, Mozambique, Cote d'Ivoire, Ethiopia, Malawi, Tanzania, and Burkina Faso, the share of older people was slightly smaller, ranging from 4 to 4.8% of the total population. The country with the lowest share of the ageing population is Uganda (3.4%), followed by Kenya (3.6%), Zambia (3.7%), and Rwanda (3.8%).

Figure 2.25 shows that older people will constitute a large share of the total population of South Africa. Projections suggest that the ageing population of South Africa will increase from 7% in 2000 to 11.5% in 2030. In 2000, the ageing population constituted 7% of the total population of South Africa, followed by Cameroon with 5.2% and Ghana with 5.1%. By 2030, projections suggest that South Africa's ageing population will form 11.5% of the total population, followed by Ghana with 9.5% and Kenya with 6.7%.

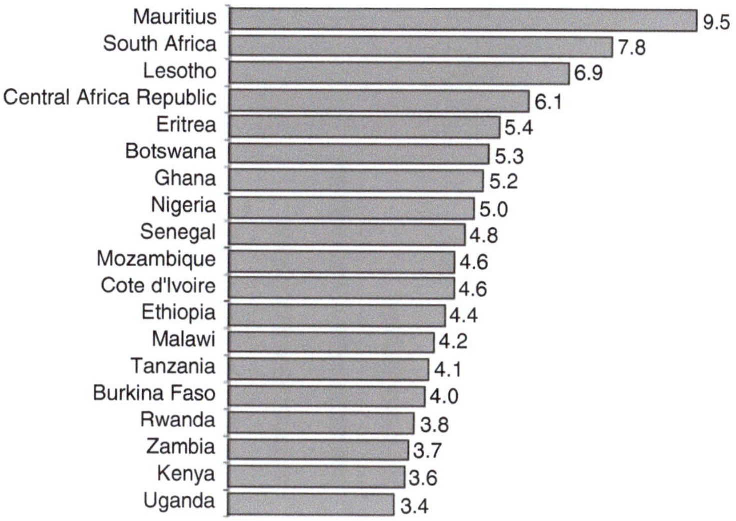

Fig. 2.24 The percentage of the ageing population in Africa countries, 2005. *Source*: Cohen and Menken (2006)

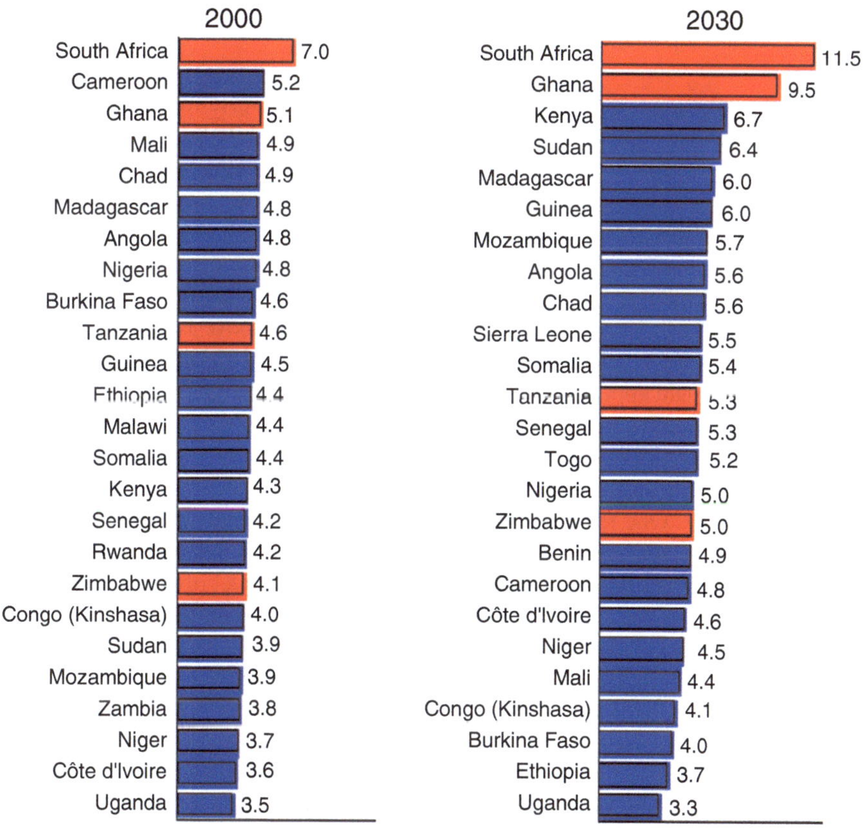

Fig. 2.25 The percentage of the population 60 years and over in African countries, 2000 and 2030. *Source*: WHO (2000)

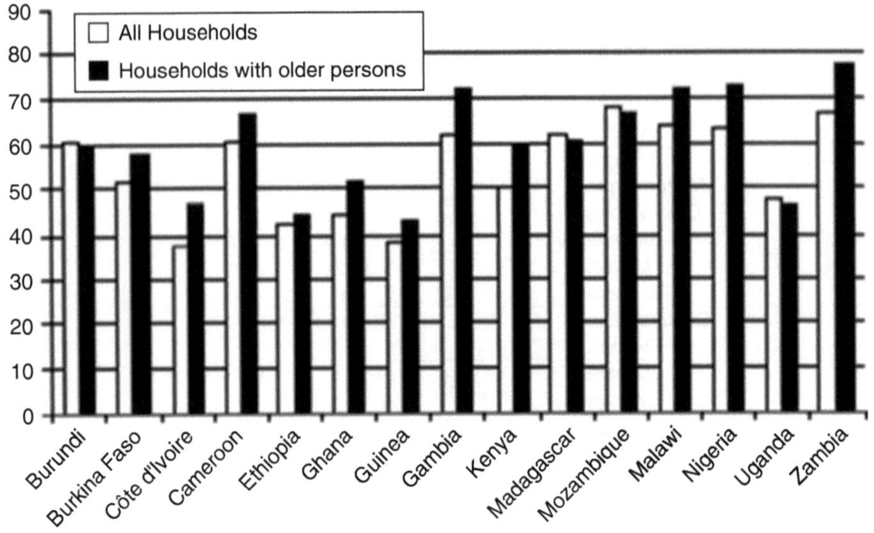

Fig. 2.26 The incidence of poverty in select African countries. *Source*: Kakwani and Subbarao (2005)

Overall, countries are not expected to experience substantial growth of the population aged 60 years and over from 2000 to 2030. While the ranking of some countries may change in relation to other African countries, the percentage of the older population is expected to grow only slightly over the 3 decades. However, there are exceptions; for instance, the older population of Sudan will grow from 3.9% in 2000 to 6.4% in 2030. High growth rates of the older population will also be seen in Mozambique, with projections suggesting an increase from 3.9% in 2000 to 5.7% in 2030. In Kenya, the population aged 60 and above is also expected to increase from 4.3% in 2000 to 6.7%, which will be the third highest in 2030. On the other hand, it would appear that the share of the older population in the total population is likely to decline in Zambia. Zambia does not feature in the chart in 2003 due to the decrease in size of its elderly population.

Figure 2.26 shows that households that include older persons are poorer than other households. One reason why older people are in a disadvantaged position is largely because household income has to be shared amongst more members, which increases the poverty rate. The countries with the highest level of poverty in households with older persons are Zambia (78%), followed by Gambia (73%), Nigeria (72%), and Malawi (71%).

The countries with the lowest level of poverty in households with older people are Guinea (42%), Ethiopia (43%), Uganda (46%), and Cote d'Ivoire (47%). The issue of high levels of poverty is exacerbated in African countries as the household income has to be shared among older as well as younger family members. Thus, the poverty rate is expected to be higher in a household that includes older people as there are more people to share the household income and resources. Unless the

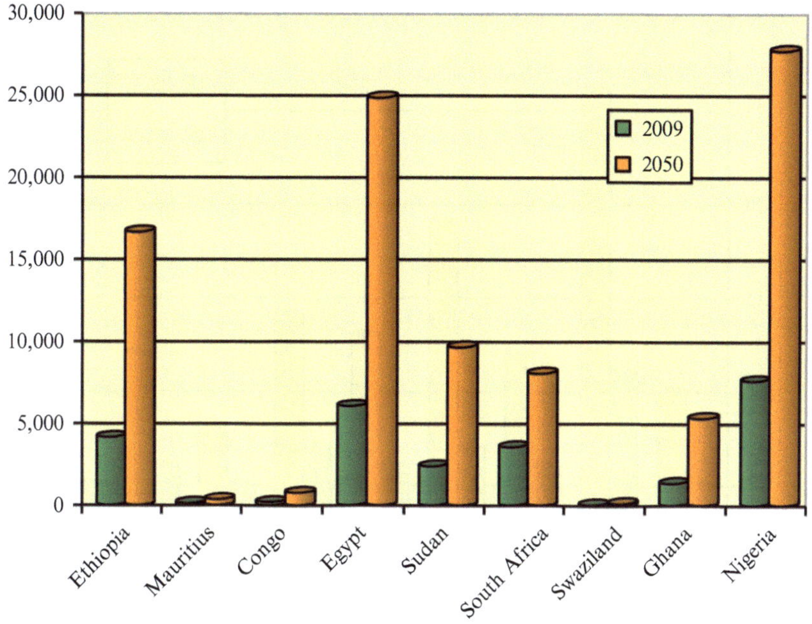

Fig. 2.27 The ageing population in thousands in select African countries, 60 years and over. *Source*: United Nations (2008a)

poverty rate in Africa is decreased to a level comparable with other continents, older people and the other household members they reside with will continue to suffer the burdens of poverty.

2.6.1 Country-Level Demographics

The following section provides an examination of the population aged 60 and over in select African countries from the five regions: Ethiopia and Mauritius in East Africa, Democratic Republic of the Congo in Central Africa, Egypt and Sudan in Northern Africa, South Africa and Swaziland in Southern Africa, and Ghana and Nigeria in West Africa. Two countries per African region were randomly selected from the United Nations World Population Prospects Comprehensive Tables (2008a). According to data from the United Nations (2008a) (Fig. 2.27), Nigeria had the highest number of people aged 60 years and over in 2009, in comparison with other select African countries, with 7,591,000, followed by Egypt with 6,054,000, and Ethiopia with 4,162,000.

In 2009, Mauritius and Swaziland had the fewest number of people aged 60 and over of the select African countries, with 145,000 and 62,000, respectively. Projections for 2050 show that if the situation remains unchanged, Nigeria will

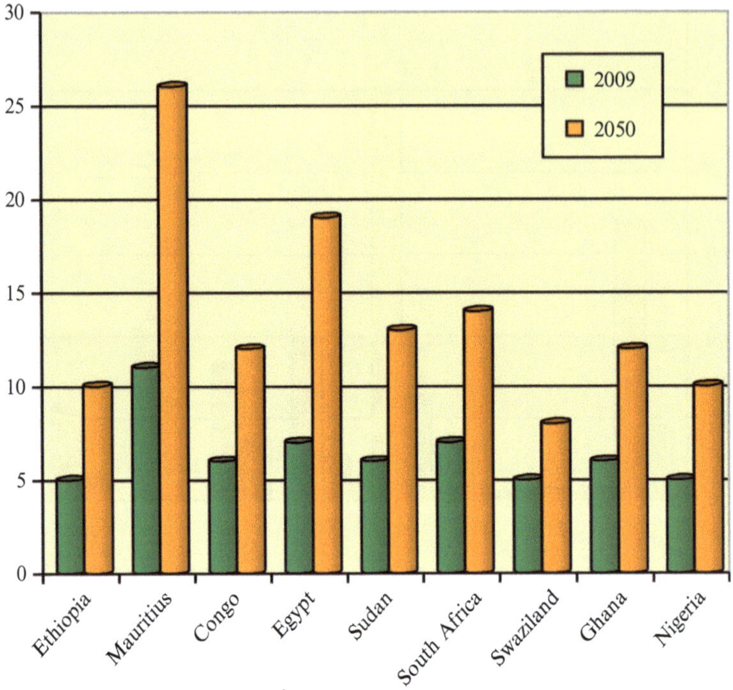

Fig. 2.28 Percentage of the total population in select African countries, 60 years and over. *Source*: United Nations (2008a)

continue to have the largest number of elderly people, at 27,719,000, followed by Egypt (24,846,000) and Ethiopia (16,658,000).

The ranking of the select countries do not alter much, as Mauritius and Swaziland will remain the countries with the lowest number of older people. It is important to point out the high numbers of older people in Northern, Western, and Eastern Africa, in comparison with the Southern and Central African countries. There is a large disparity in the size of the ageing population by region.

Although Mauritius has the second lowest number of older men and women (Fig. 2.27), when the population aged 60 and over are considered as a percentage of the total population (Fig. 2.28), the country has the largest share of the older population. Mauritius ranks the highest of all the countries with 11% of older people contributing to its total population in 2009. The share of elderly in the total population of the other countries ranges between 5 and 7%.

Projections for 2050 indicate that Mauritius will remain the leading country with the highest percentage of elderly in the total population. However, the percentage of elderly in Mauritius will rise to 26%, followed by Egypt with 19%, and South Africa with 14%.

In the remaining countries, the share of the elderly as a proportion of the total population will range from 8 to 13%. It is important to note that, other than Mauritius

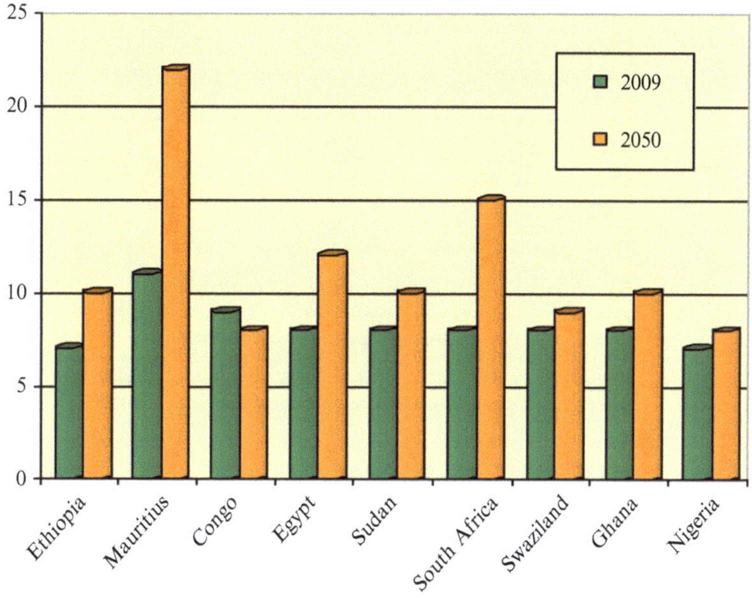

Fig. 2.29 Share of older persons in select African countries, 80 years and over. *Source*: United Nations (2008a)

with an elderly population that forms one fifth of the total population, all the other countries have a percentage of elderly people who form less than a fifth of the total population. Over the next few decades the ageing population is expected to grow, but the population will continue to consist mainly of the youth and the economically active age group. This validates the future demographic trend of a "young" African population majority.

Figure 2.29 illustrates the percentage share of persons aged 80 years and over in the total population. Mauritius has the highest share of people in the oldest age cohort (11%), followed by Congo (9%), then Egypt, Sudan, South Africa, Swaziland, and Ghana, all with 8%. The share of elderly above the age of 80 years per country ranges from 7 to 11% in 2009.

Projections for 2050 indicate that Mauritius will continue to have the largest share of the population aged 80 years and over, with 22%. South Africa will have the second largest share with 15%, followed by Egypt with 12%. The oldest cohort in the remaining countries ranges from 8 to 10%. Other than Mauritius, in these select African countries, the oldest cohort comprises less than a fifth of the total population.

Figure 2.30 illustrates the sex ratio of the ageing population for select African countries, with Ghana as the leading country for both age groups: 60 years and over and 80 years and over. An examination of the population aged 60 years and over shows that Ghana has 94 men for every 100 women, followed by Sudan with 88

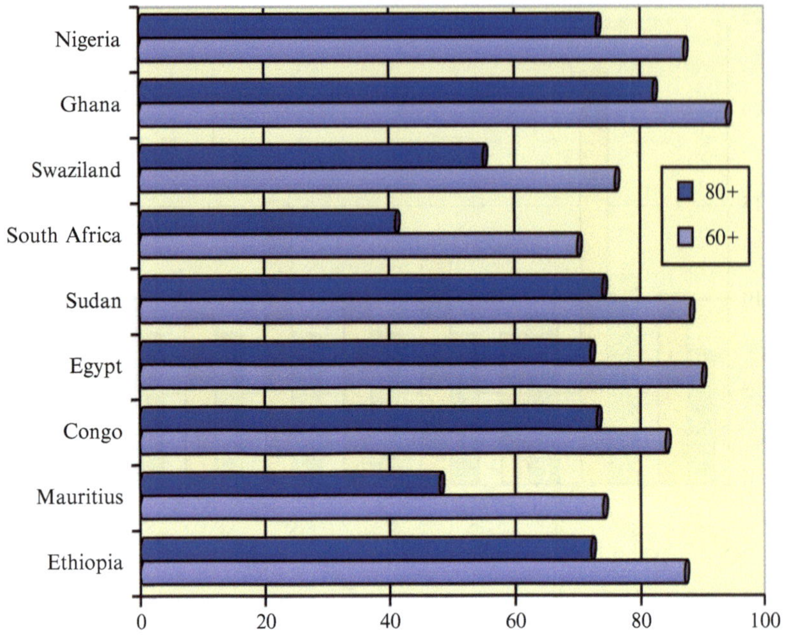

Fig. 2.30 The sex ratio of the ageing population in select African countries, 2009. *Source*: United Nations (2008b)

men for every 100 women, and Congo and Nigeria with 87 for every 100 women, respectively. These are relatively high sex ratios, which indicate that there are almost as many men as there are women in the population aged 60 years and over. The sex ratios for Mauritius and South Africa vary somewhat from other countries, with less than 75 men for every 100 women aged 60 years and over.

An examination of the sex ratios of the oldest age cohort in the population shows that Ghana has the highest ratio of men to women. There are 82 men for every 100 women aged 80 years and over. Thus, in Ghana, in the oldest cohort there are almost as many men as women. This is followed closely by Sudan with 74 men per 100 women, and Nigeria and Congo with 73 men per 100 women. All the other African countries have a sex ratio above 50 men per 100 women, with the lowest in South Africa with 41 men per 100 women.

2.6.2 Gender Differentials

Figure 2.31 shows that the life expectancy of the selected countries is similar overall, at a country level, and for both men and women. In all the countries, women are expected to outlive men. The country with the highest life expectancy at age 60 for men is Mauritius, with men living an additional 17 years after reaching the age of

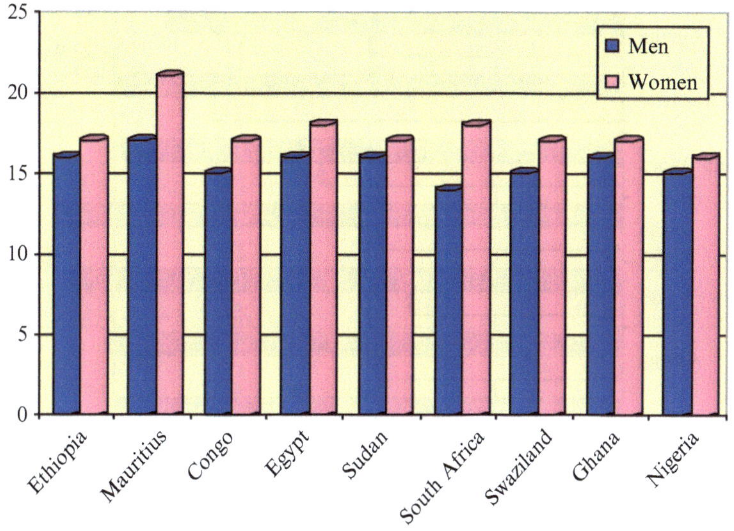

Fig. 2.31 Life expectancy at age 60 in select African countries, 2005–2010. *Source*: United Nations (2008b, c)

60. Women in Mauritius also experience the highest life expectancy at age 60, with women living an additional 21 years after reaching the age 60. South Africa has the second highest life expectancy of all selected African countries, with women expected to live an additional 18 years after reaching the age of 60.

The life expectancy for men at 60 years in these African countries ranges from 14 to 17 years, and for women ranges from 16 to 21 years. Thus, women have a longer life expectancy at age 60 than men. However, although African women may have a longer life expectancy than African men, the downside is that they may have to contend with greater ill-health in their lifetime. In addition, they are more likely to be alone as they may have outlived their male partners.

According to Fig. 2.32, in all the selected African countries, a higher percentage of older men are married than women. The greatest percentage of older, married men is found in Nigeria and Sudan with 89%, followed by Ethiopia with 87%. Older women are less likely to be married than older men, and the country with the highest percentage of married is Swaziland, with 64% of older, married women.

Figure 2.33 shows that men in select African countries are expected to have greater companionship in their old age than women due to their greater propensity to be married in their old age.

In five of the seven select countries (Fig. 2.33), there are more older women living alone than older men, with Egypt and Ghana taking the lead with 16% each. This is in stark contrast to South Africa and Swaziland. In South Africa a similar proportion of older men and women live alone, at 8%. In Swaziland, more older men than women live alone. In Swaziland 9% of older men live alone, compared with 6% of older women.

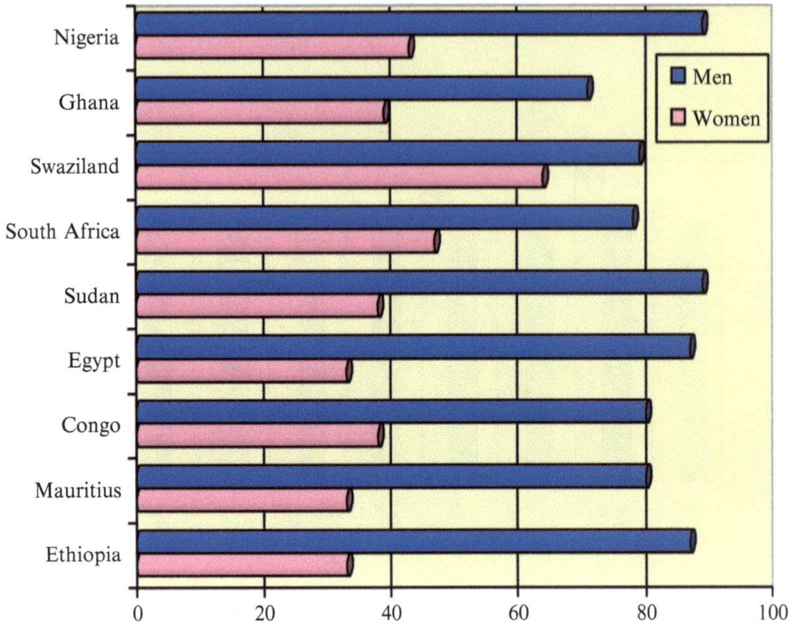

Fig. 2.32 Percentage married in select African countries, 60 years and over ("married" includes those in consensual unions). *Source*: United Nations (2008b, c)

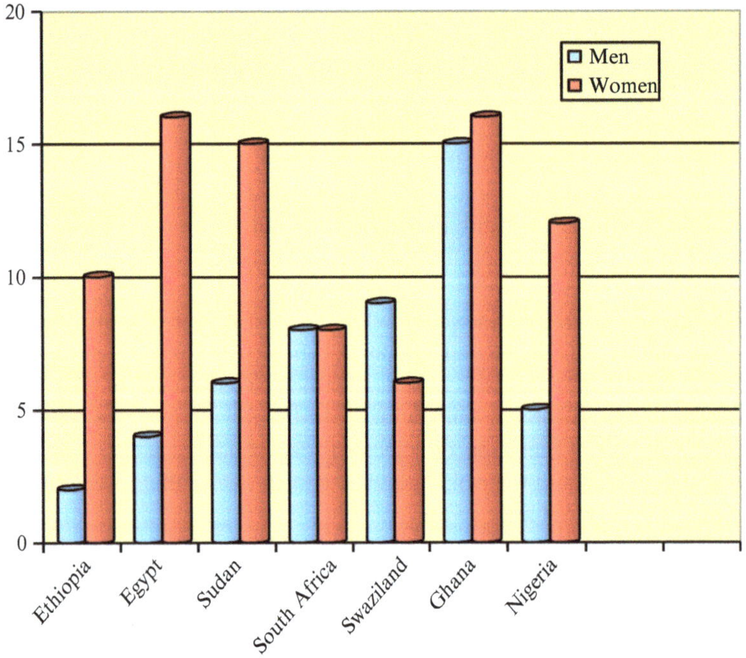

Fig. 2.33 Percentage living alone in select African countries, 60 years and over. *Data not available for Congo and Mauritius. *Source*: United Nations (2008a, b)

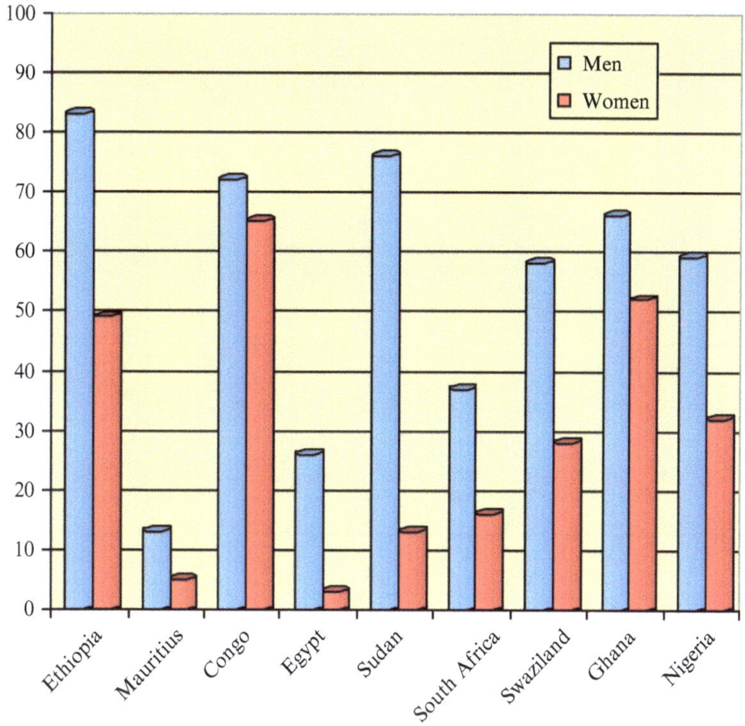

Fig. 2.34 Percentage in the labour force in select African countries, 60 years and over. *Source*: United Nations (2008b, c)

What is striking in selected African countries is the disparity between older men and women who live alone. For instance, in Egypt, older women (16%) are four times more likely than older men (4%) to live alone. In Ethiopia, older women (10%) are five times more likely than older men (2%) to live alone, and in Sudan and Nigeria more than twice as many older women live alone than men. Thus older women are in a far less advantageous position than older men in African countries as they are more likely to lack companionship in their old age. This highlights the issue of gender inequality in select African countries.

In all the select African countries men aged 60 and over are more economically active than women (Fig. 2.34). In Ethiopia, men aged 60 and over are the most economically active with 83% in the labour force. The percentage of economically active older men ranges from 13% in Mauritius to 83% in Ethiopia.

Congo has the most economically active women aged 60 and over, with 65% in the labour force, followed by Ghana with 52%, and Ethiopia with 49%. The proportion of older women in select African countries participating in the labour force ranges from 3% in Egypt to 65% in Congo. The greatest disparity in the labour force participation of men and women is seen in Sudan with 76% of older men still part of the labour force, compared with only 13% of older women.

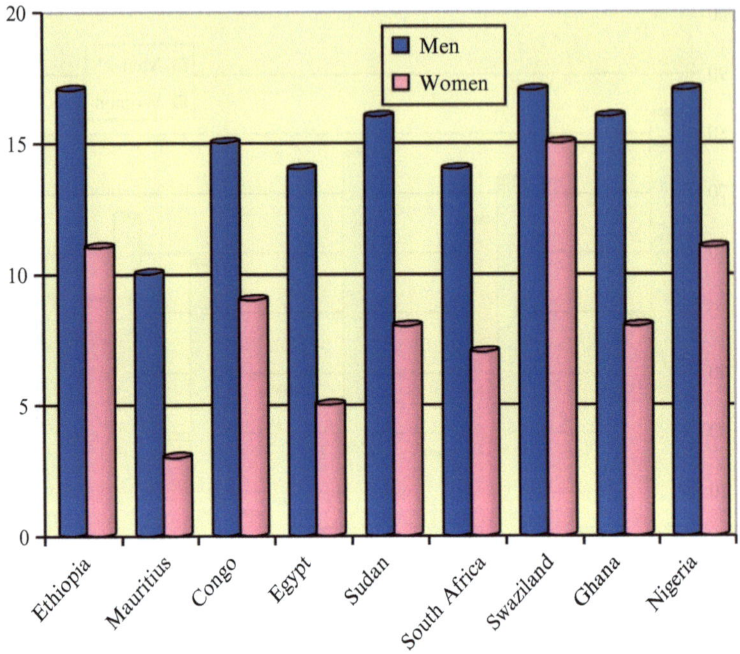

Fig. 2.35 Old age support ratio in select African countries, 2009. *Source*: United Nations 2008b

Figure 2.35 illustrates the support ratio of older men and women, with older men having more support from the economically active population. Comparison shows that the countries have a similar old age support ratio. For each older man in Ethiopia, Swaziland, and Nigeria, there are 17 economically active persons to support them. In Swaziland, older women have almost as much support as men, with 15 economically active persons per older woman.

However, older women from the other select African countries do not receive as much support as older men. The greatest disparity is seen in Mauritius with a support ratio of 10 for older men and 3 for older women. In the select African countries, men receive far greater support than women from the economically active.

2.7 Conclusion

Over the next 4 decades Africa will experience exponential growth of its ageing population. The ageing population of Africa is growing at a much faster pace than any other continent. By the year 2050, the population aged 60 years and over will comprise 10.4% of the total African population (United Nations 2007). Thus, more effort should be focused on population ageing, since the elderly will constitute a substantial share of the total population of Africa in the near future.

The profile of the ageing population in Africa shows that the life expectancy is expected to increase steadily. This is consistent with the argument by Van Dulleman (2006), who observed that the impact of HIV and AIDS and other factors which decrease population size will not offset the growth of the elderly population. Furthermore, there will be more elderly women in Africa than men. In other words, older women will have a higher life expectancy than men. The greater life expectancy for both older men and women necessitates greater social protection for them in the form of old age pensions, health services, and housing.

In terms of geographical location, older people are evenly dispersed between rural and urban areas. This is a population trend that will continue into the future, as projections indicate that by 2017, the number of urban dwellers will equal rural dwellers (Gündel 2006). An examination of gender differentials indicate that older men are more likely than older women to be married in Africa which have implications for their health. Studies suggest that the married have better health than the unmarried (Waite 1995). Furthermore, economic data indicate that men dominate the older workforce, with more men participating in the labour force at older ages than women. In addition to this, older men have a further advantage over women as they encounter greater support from the economically active population. Thus, elderly men are more likely to have a better socio-economic status than women. The feminisation of the older population and the prevailing gender inequalities are likely to act as a barrier to the growth and success of the continent, therefore, structured efforts must be made to address this issue.

In terms of educational status a little more than half of all older Africans are literate; however, despite their lack of education in comparison with older people in other continents, they remain the most economically active ageing population in the world. The high level of economic activity among the population aged 60 and over is not surprising, given the lack of social protection schemes in many African countries.

The regional trends in ageing have shown that Northern Africa will experience the largest growth of its ageing population over the upcoming decades. Projections indicate that by 2050, Northern Africa's ageing population will be four times larger than today. Life expectancy will increase in all African regions. Thus, these regional trends indicate that adequate focus must be placed on the elderly in Northern Africa especially, as this region will constitute a sizeable portion of the older African population, due to population growth and increasing longevity, respectively. The increase in the older population calls for an improvement in the health systems of African nations. However, there is a recognition that an improvement in African health systems remains challenging, considering that the continent is home to the worst health systems in the world (Robinson 2007). Nevertheless, it is vital that national governments in Africa take the necessary steps to address deficiencies in their health system in order to promote a supportive and sustainable future for older Africans.

References

Apt, N. (2000). *Rapid urbanization and living arrangements of older persons in Africa*. New York: United Nations Department of Economic and Social Affairs.

Cohen, B., & Menken, J. (2006). *Ageing in Sub-Saharan Africa: Recommendations for Furthering Research*. Washington DC: The National Academies Press.

European Centre for Social Welfare Policy and Research. (2008). *Features and challenges of population ageing: The European perspective*. Vienna: European Centre.

Gelb, A. (2001). Gender and growth: Africa's missed potential. *Development Outreach* 3, 2.

Gündel, S. (2006). *A synthesis of urban and peri-urban agricultural research commissioned by the RNRRS programme. Renewable natural resources research strategy*. Edinburgh: RNRRS.

HelpAge International. (2008). *Mind the gap: HIV and AIDS and older people in Africa*. London: HelpAge International.

Kakwani, N., & Subbarao, K. (2005). *Aging and poverty in Africa and the role of social pensions. International Poverty Centre, Working Paper No. 8*. New York: United Nations Development Programme.

Kalasa, B. (2004). *Population ageing in Africa: A Policy Dilemma*. Addis Ababa: UNFPA.

Kinsella, K., & Wan, H. (2009). *An ageing World: 2008*. Washington DC: US Census Bureau.

Knodel, J., & Ofstedal, M. B. (2003). Gender and ageing in the developing World: where are the men? *Population and Development Review, 29*(4), 677–698.

Lam, D., Leibbrandt, M., & Ranchhod, V. (2006). Labor force withdrawal of the elderly in South Africa. In B. Cohen & J. Menken (Eds.), *Ageing in Sub-Saharan Africa: Recommendations for furthering research*. Washington DC: The National Academies Press.

Lesthaeghe, R. (2000). *Europe's demographic issues: Fertility, household formation and replacement migration* (United Nations Expert Group meeting on policy responses to population ageing and population decline). New York: United Nations.

Lloyd-Sherlock, P. (2000). Population ageing in developed and developing regions: implications for health policy. *Social Science and Medicine, 51*, 887–895.

Robinson, M. (2007). *Global health and global ageing*. San Francisco: Jossey-Bass.

Rwezaura, B. A. (1989). Changing community obligations to the elderly in contemporary Africa. *Journal of Social Development in Africa, 4*(1), 5–24.

UNDP. (2005). *The diversity of changing population age structures in the World*. Argentina: Paulo Saad.

UNFPA. (2008a). *Population ageing and development: Social, health and gender issues* (Population and development strategies). New York: United Nations Population Fund.

UNFPA. (2008b). *2008 Revision of World population prospects*. New York: United Nations Population Fund.

United Nations. (2001). *World population ageing: 1950–2050*. New York: United Nations Department of Economic and Social Affairs.

United Nations. (2004). *World population prospects: The 2004 revision*. New York: United Nations Department of Economic and Social Affairs.

United Nations. (2007). *World population prospects: The 2006 revision*. New York: Department of Economic and Social Affairs.

United Nations. (2008a). *World population prospects: The 2008 revision. Vol. I: Comprehensive tables, United Nations Publication: ST/ESA/SER.A/287*. New York: United Nations.

United Nations. (2008b). *World population prospects: The 2008 revision. Vol. II: Sex and age distribution of populations. United Nations Publication: ST/ESA/SER.A/291*. New York: United Nations.

United Nations. (2008c). *World population prospects: The 2008 revision. Vol. III: Living arrangements of older persons around the World. United Nations Publication: ST/ESA/SER.A/240*. New York: United Nations.

United Nations. (2009). *World population ageing 2009*. New York: Department of Economic and Social Affairs: Population Division, United Nations Publication.

United Nations. (2010). *Current status of the social situation, well-being, participation in development and rights of older people worldwide*. New York: Department of Economic and Social Affairs. Office of the Commissioner for Human Rights.

United Nations. (2011). *World population prospects: The 2010 revision*. New York: Department of Economic and Social Affairs, Population Division.

Van Dulleman, C. (2006). Older people in Africa: new engines to society? *Feminist Formations, 18*(1), 99–105.

Waite, L. J. (1995). Does marriage matter? *Demography, 32*(4), 483–507.

WHO. (2000). *Information needs for research, policy and action on ageing and older adults*. Bethesda, MD: U.S. National Institute on Ageing.

Chapter 3
The Health Situation of Older People in Africa

Farzana Alli and Pranitha Maharaj

3.1 Introduction

Over the past decade, a changing population age structure has contributed to an increase in the proportion of elderly people in Africa. For developing regions, population ageing raises a serious health concern as many chronic illnesses, such as cardiovascular disease, respiratory illness, and diabetes increase with age. In addition to the impact of chronic illnesses on the social and economic situation of the affected elderly, a change in the disease profile will have considerable implications for health service provision. Providing insight into the health and well-being of Africa's older population can help guide policy development and strengthen institutions at the local, national, and international level to respond effectively to population ageing. This chapter will profile the health situation of the older population in Africa. It will focus on the prevalence and nature of illnesses in the older population in selected African countries. The chapter will also examine the leading causes of morbidity and mortality in Africa among the elderly, and focus more specifically on the chronic disease burden. Obtaining data on the health situation of the older population in Africa is often difficult. Despite this, every effort will be made to review the situation in a wide range of countries in Africa as possible.

F. Alli (✉)
Health Economics and HIV/AIDS Research, University of KwaZulu-Natal,
Durban, South Africa
e-mail: alli.farzana@gmail.com

P. Maharaj
School of Built Environment and Development Studies, University of KwaZulu-Natal,
Durban, South Africa
e-mail: Maharajp7@ukzn.ac.za

P. Maharaj (ed.), *Aging and Health in Africa*, International Perspectives on Aging 4,
DOI 10.1007/978-1-4419-8357-2_3,

3.2 The Definition of Health and Well-Being

Over the last few decades, countries around the world have advocated for health as a fundamental human right to be attained by all people. However what constitutes health and well-being differs significantly across cultures, between individuals, and over social classes (Robinson et al. 2007). Though the issue of health and disease has been a major concern since ancient times, the use of the word "health" to describe "well-being" is relatively recent (Awofeso 2010). Prior to the first century, health was perceived as a divine gift while ill health was viewed as a punishment inflicted by the gods. Hippocrates, considered one of the most influential figures in the history of medicine, essentially pioneered the move away from divine notions of health by making use of observations as a basis to acquiring health knowledge (Awofeso 2010). Nonetheless, centuries later, a divine view of health persists particularly within Africa (Truter 2007; Awofeso 2010). In this context, health is often intertwined with cultural and religious beliefs. This definition does not only focus on the physical condition but also the social, psychological, and spiritual aspects (Truter 2007).

In the twenty-first century, one of the most common definitions of health is that formalised by the World Health Organisation (WHO) which classifies health as being "a state of complete physical, mental, and social well-being and not merely the absence of disease or infirmity". Though this definition was conceived over 50 years ago, it is still widely accepted and relevant today in informing health policy and practice (Robinson et al. 2007). It suggests a more holistic approach which views health as comprising a variety of factors including an individual's physical, social, and mental health. While physical health refers to the body's ability to function, mental health is defined as a "state of well-being in which an individual realises his or her own abilities, can cope with the normal stresses of life, can work productively and is able to make a contribution to his or her community" (WHO 2003a). From this definition, mental health forms the basis for individual well-being and the effective functioning of a community (WHO 2003a). Social health and well-being refers to the way in which individuals interact with people in their environment. Multiple social, psychological, and biological factors can influence the mental health of a person. Often referred to as the health triangle, this model suggests that a person is only healthy if these three elements exist. Evidence shows that physical health affects psychological health, while research on somatic disorders suggests that psychological well-being similarly has an impact on physical health (WHO 2003a). At a broader level, social well-being has been shown to have a preventative effect on both physical and psychological ill health, which can be defined as the "social experience of being sick or diseased" (Lamb and Siegal 2004, p. 342).

In recent years, critics have argued that the WHO definition of health is inflexible and unrealistic. They argue that the use of the word "complete" in the definition makes it highly unlikely that anyone would be healthy for a reasonable period of time (Awofeso 2010). The statement "a state of complete physical, mental, and social well-being" is also associated more with happiness than health. Awofeso (2010) therefore argues that different definitions that take into account changing health needs should be used to revise and improve on the current WHO definition of health. In the context of Africa this is important since health is perceived in different ways across cultures, between individuals, and over social classes. Nonetheless, based on its holistic view of health, this model forms a good basis for understanding health in the African context.

3.3 Why Focus on the Health of the Elderly?

Over the past decade, the world population has increasingly been characterised by a rapidly growing older population. Demographic trends globally and in Africa reveal that there has been a considerable rise in the population aged 60 and over, driven by declining rates of fertility and mortality (Kowal et al. 2010). The older population in Sub-Saharan African continues to grow more rapidly than in the developed world, at an average annual rate of increase of 2%. This rate is projected to increase to nearly 4% over the next 50 years (Velkoff and Kowal 2006).

Population ageing has a profound impact on the economic and health systems of countries including the ability to maintain contributions and provide resources for older populations (Kowal et al. 2010). Traditionally, in developing countries, older people were primarily supported by the household and family, supplemented in many instances by informal mechanisms including mutual aid societies and kinship networks (Cohen and Menken 2006; WHO 2003a). Despite a dearth of empirical research on the long-term welfare of the elderly, studies are suggesting that traditional caring and social support systems are becoming increasingly strained (Robinson et al. 2007). Increasing development and modernisation, together with social and economic changes, has resulted in the weakening of traditional values and networks. The modernisation of societies has been associated with several crucial issues which affect the family support of older people:

- Low fertility, which results in fewer children able to provide familial support
- Increased female labour force participation, which decreases the number of caregivers who can provide support to older people
- Formal education, which has been linked to weakening traditional social ties and responsibilities regarding familial support
- Rural–urban migration, which results in young people seeking employment in urban areas, leaving behind older family members in rural areas (World Bank 2007)

In most African countries, formal pensions and other social welfare schemes are virtually non-existent. Where they do exist, limited resources have led to minimal packages for basic health services for the elderly. Consequently, older people are faced with a heavy burden of health issues without adequate social security systems or well-functioning traditional care systems. With a reduced ability to generate resources, older people lack basic needs that affect their health status (Waweru et al. 2003).

The direct impact of an ageing population on a country's economy has also received increasing attention in recent years. In coming years, with a changing age structure, the elderly will make up a substantial proportion of the workforce (Robson 2001). One of the factors that will contribute to this change in the profile of the workforce is the projected labour force shortage (Pandey 2009). When the labour supply starts to diminish, employers will seek non-traditional sources such as older workers to overcome the labour shortage (Lockwood 2003). Studies show that health has a significant and positive effect on the participation of the elderly in the workforce (Pandey 2009). Poor health among the older workforce may lead to low participation rates and impact negatively on the performance of the economy (Lockwood 2003). A better understanding of the health situation of the older population will help inform economic policy linked to labour market participation.

Ageing populations have been associated with rising demands on the health care system stemming from a change in the disease profile. With demographic ageing, the prominent causes of death change from those associated with infant and child mortality to those associated with old age (Joubert and Bradshaw 2006). Similarly the disease burden shifts from infectious, maternal, and nutritional conditions to chronic, non-communicable, and degenerative diseases (Joubert and Bradshaw 2006). This transition to higher levels of chronic diseases among the older population will demand a shift to more self-management, home-based care, and continuous care, which has serious implications for the physical and human infrastructural aspects of health (Robinson et al. 2007). Due to the current configuration of health systems, most public health programmes are directed at eradicating or controlling preventable childhood diseases rather than treating or managing chronic diseases and the health care of the elderly (Cohen and Menken 2006). Many African health systems are facing budget constraints, and are struggling to cope with the cumulative burden of communicable and chronic diseases. Over the last decade, an estimated 80% of regional health budgets were allocated to infectious diseases while a smaller fraction of health budgets were spent on treating illness among older adults (de-Graft Aikins et al. 2010; Mwanyangala et al. 2010). Due to social and financial barriers, access to health care for the elderly is also limited and not a major priority in most developing countries. A shift to non-communicable and chronic diseases will thus challenge health systems without the necessary data essential to informing policy and planning (Kowal et al. 2010).

Though worldwide improvements in health have contributed to an increase in longevity, developing countries continue to face numerous challenges in improving the health and well-being of their older population. In the past, the main health problems faced in developing countries were communicable or infectious diseases. However, fuelled by population ageing, an "epidemiological revolution" has seen the disease burden shift from communicable to non-communicable diseases (Robinson et al. 2007). African countries are now faced with a dual burden of infectious and chronic diseases. Existing research suggests that infectious diseases account for at least 69% of deaths on the continent while age-specific mortality rates from chronic diseases are higher in Sub-Saharan Africa than in virtually all other regions of the world (de-Graft Aikins et al. 2010). According to the WHO (2003a), the burden of diseases at older ages is characterised by high levels of non-communicable diseases, increased levels of chronic diseases, as well as high rates of injuries and violence. Studies from Sub-Saharan Africa show that some of the prominent non-communicable diseases among the older population are hypertension, heart disease, diabetes, and stroke (Aboderin 2008). Other non-communicable diseases include musculo-skeletal conditions, visual impairments, and mental disorders. Based on the experience of industrialised nations, the projected increase in non-communicable diseases will be accompanied by growing levels of disability and an increasing loss of physical and cognitive functioning (Suzman 2010). In Africa, the burden of chronic non-communicable disease is expected to increase from 28 to 51% over the next 20 years (Nawi et al. 2010). Studies predict that the dominant scenario in many developing countries will be co-existing chronic infections and non-communicable diseases (Nawi et al. 2010). For most of the developing world this will be compounded by high levels of poverty and poor sanitation, along with diseases such as malaria, tuberculosis, and AIDS (Robinson et al. 2007). Epidemiological studies conducted in this context suggest that population ageing will have a significant impact on global health. It is projected that by 2020 almost three quarters of all deaths in the developing world will result from old age (Lloyd-Sherlock 2005).

Focusing on the health and well-being of the elderly is important for informing policy and planning. In many Sub-Saharan African countries, the health of older people is linked to broader socio-economic and cultural factors. In countries that are severely afflicted by HIV and AIDS, changing household structures have had significant implications for the roles and responsibilities of older people (Zimmer and Dayton 2005). The availability of highly active anti-retroviral therapy has meant that AIDS is becoming a chronic disease requiring long-term care. Older people, particularly women, are burdened with the responsibility of caring for sick and surviving children and relatives, as well as assuming childrearing responsibilities for their grandchildren (Hosegood and Timaeus 2006). With the vast majority of individuals living with HIV in their prime wage-earning years, their principal sources of financial and material support have diminished (Robinson et al. 2007). Increasingly, as a consequence of AIDS, older people are given the responsibility of providing emotional and economic support, primarily through their social pensions and old-age grants, to both their own children and grandchildren (Gomez-Olive et al. 2010; Hosegood and Timaeus 2006). This has significant implications for the health and well-being of the elderly. A case study on elderly caregivers conducted in Zimbabwe has shown that the physical and emotional well-being of the elderly is compromised by the daily responsibilities and duties of providing care for AIDS patients and children left orphaned by the disease (WHO 2002). The study also documented a high prevalence of verbal abuse, stigmatisation including accusations of witchcraft, and in some cases physical violence directed at the caregivers (WHO 2002). Studies have concluded that the stress of providing care to AIDS patients is therefore directly associated with greater symptoms of depression and poor health among the elderly (Joubert and Bradshaw 2006). In addition, older people are also directly affected by HIV/AIDS. Studies show that an estimated 3 million elderly individuals in Africa are living with the virus accounting for approximately 14% of all HIV infections (Mills et al. 2011). Most treatment and intervention programmes focus on the needs of young children and middle-aged adults; African societies therefore face a unique challenge in delivering services orientated towards the circumstances and needs of older people (Economic Commission for Africa 2007). Apart from the impact of AIDS, research shows that older people are at an increased risk for chronic disease, disability, injury, and diminished well-being (Joubert and Bradshaw 2006). The ageing process is frequently accompanied by declining physical capacity, decreased sensory abilities such as vision and hearing, and changes in mental functioning. The AIDS epidemic has thus significantly increased the vulnerability of older people at a time in their lives when they most likely require care and support.

For these reasons, the health of the older population has become a prominent issue that may impact the well-being of the entire population. Understanding their health needs can provide significant insights into the complexities of ageing, which will enable governments to address the vulnerabilities of older people who are playing a key role in providing care and support to households afflicted by AIDS (WHO 2002). At the same time, it is essential for informing policy design and implementation, particularly with regard to social security systems and resource distribution. Despite the evident need to understand issues that affect the health of the older population, only recently has this become the focus of attention in Africa. As such, the health needs of the older population as well as the implications of national policies for health and welfare are poorly understood. It is projected that Africa will experience the fastest rate of growth in the numbers of older people compared with any other continent (Economic Commission for Africa 2007). The older population in this context is expected to reach 64.5 million by 2015 and exceed 103 million in 2030. This unprecedented demographic shift has serious implications for the epidemiological profile of the population (Gomez-Olive et al. 2010). As the population grows older and there is an increase in the number of chronic conditions, the demand for social support and health services is also expected to increase significantly over the next few years. Though there are significant differences in the health challenges facing Africa compared to higher-income countries, over the last century, the health needs of older people have been marginalised or excluded from international policy initiatives applied in the developing world (Cohen and Menken 2006; Lloyd-Sherlock 2005). Currently, health systems in the continent are geared towards meeting the needs of women and children, with little or no focus on the elderly. With non-communicable diseases reaching epidemic proportions, health systems will need to be re-orientated towards addressing the health needs of ageing populations.

3.4 Challenges of Measuring Health in Africa

One of the main limitations of studying the health of older people in Africa stems from the lack of research on ageing. This is largely influenced by the availability and quality of data on the older population. An important source of health data in many developing countries is the Demographic and Health Survey but it usually excludes the older population. Existing data on older people is obtained primarily from censuses which are not always reliable or detailed and small cross-sectional surveys which cannot always be generalised to the entire population (Cohen and Menken 2006). Micro-level data is available for only a limited number of countries. In addition there is a scarcity of longitudinal data which is essential in understanding many of the dynamics of ageing over a period of time (Suzman 2010). In many African countries there is a dearth of reliable information on health risks, disease, and disability in the older population (Joubert and Bradshaw 2006). Vital statistics and administrative records are often inadequate and incomplete in developing countries. For instance, fewer than ten countries have vital registration systems that produce adequate data while only few systems in countries including Mauritius, Seychelles, and South Africa cover at least 80% of the population (Velkoff and Kowal 2006).

Health information is gathered through self-reported health status which is a common measure of health in Africa. Studies suggest that self-reported health status predicts morbidity and subsequent mortality, and allows for the examination of how health status varies over the life course (Pandey 2009). However, many studies have questioned the reliability of self-reported data. Some concerns include the over-reporting of health problems or under-reporting of positive health status (Pandey 2009). A common problem is the discrepancy between objective indicators of health and subjective reports (Lamb and Siegal 2004). In many Sub-Saharan African countries individuals' understanding of illness may also differ. In addition, data on health conditions is often obtained through probes for specific conditions and, as a result, certain health issues may not be documented (Lamb and Siegal 2004). Differences in the way people define old age and households in Africa may also contribute to problems in measuring health. For instance, when participants in a focus group interview in Nairobi were asked to describe who they perceive as old, they claimed that they can be identified by their physical characteristics, life experiences, or social roles (Cohen and Menken 2006). According to their definition, men and women younger than 60 may be defined as old because they exhibit certain characteristics or take on social roles that are associated with the older population. Similarly, in Africa the definition of a household is problematic since many societies are characterised by extended family systems or multigenerational households (Cohen and Menken 2006). These issues complicate the study of older people's health and well-being.

Some of the key challenges in facilitating research in the Sub-Saharan African context include a lack of access to funding, scarcity of highly skilled local researchers, and restrictive administrative barriers to conducting research (Suzman 2010; Cohen and Menken 2006). One of the suggestions for improving the data is the use of multidisciplinary research designs to enhance understanding of the ageing process (Cohen and Menken 2006). Longitudinal studies are useful but expensive. One suggestion by Cohen and Menken (2006) is to use existing data collection efforts (such as the International Network for the Demographic Evaluation of Populations and Their Health (INDEPTH) sites) to obtain more information on older people. INDEPTH sites have relatively large populations under surveillance, and they are involved in regular monitoring of vital events (Boerma 2010). Their socio-demographic surveillance forms an important basis for understanding health in developing countries, particularly within rural areas (Suzman 2010). In addition, they have established training centres for health and demographic surveillance systems which provide for the training of scientists in low- and middle-income countries on complex longitudinal data analysis techniques (Sankoh 2010). Over the past several years, international institutions such as the United States National Institute on Ageing have encouraged efforts to develop nationally representative studies on adult health and ageing in low-resource countries (Suzman 2010). The WHO Study on Global AGEing and Adult Health (SAGE) is one of the first cross-national surveys focusing on ageing in Africa. These studies aim for a longitudinal cohort design with the inclusion of populations aged 50 and over, as well as a comparative cohort of people aged 18–49 (Boerma 2010). Developing countries which lack data on ageing should therefore make a determined effort to utilise this research for informing policy on ageing and health.

Qualitative studies may also make contributions to research on ageing particularly in settings where no data exists. Studies show that qualitative research on the ageing population is important since the determinants of ageing are multifaceted and complex (Liehr and LoBiondo-Wood 2006). A qualitative component can provide detailed descriptions and insights into individuals and processes that quantitative data alone may not generate (Curry et al. 2009; Liehr and LoBiondo-Wood 2006). In addition, it may offer other benefits by including individuals that are traditionally underrepresented in research, and those with low literacy (Curry et al. 2009). In Africa, with high levels of illiteracy and low levels of education, qualitative data collection methods such as focus groups and interviews may be more effective and less intimidating than quantitative methods (Curry et al. 2009). Also, given that there is little consensus on the definition of old age, qualitative approaches may be beneficial in understanding the socio-cultural factors that influence health and ageing. For developing countries which are often faced with resource limitations, conflict situations, and political instability, collection of quantitative data is often problematic, so qualitative data may be particularly useful.

Information that is relevant and specific to the older population and their situation is a priority area. Improving the quality and availability of data will assist in improving our understanding of the ageing process in Africa. However, achieving high-quality data requires sustained effort and support from governments in prioritising ageing issues on national agendas and further developing local research capacity.

3.5 Myths About the Elderly

In Africa, older people have traditionally been highly valued for their wisdom and knowledge, their position as the head of the family, and their roles in conflict resolution (Kyobutungi et al. 2010). However, a number of myths prevailing about the elderly have served to undermine their position in society. One of the most common associations is that of older people, especially women, to witches or witchcraft. The ageing process which is often accompanied by changing physical appearances may lead to stigmatisation, assault, and, in the most extreme cases, death (de-Graft Aikins et al. 2010). Myths around the health of the elderly are based on the African perceptions of health which are often intertwined with cultural and religious beliefs. Several studies in this context have shown that health issues facing the elderly are often interpreted in non-medical terms (Lloyd-Sherlock 2005). Information on the nature and occurrence of mental disorders are especially limited in Africa. In some countries, mental health disorders are perceived as bewitchment (Ferreira and Makoni 2002 cited in Lloyd-Sherlock 2005). An ethnographic study conducted in Ghana found that, in contexts where chronically mentally ill people had no access to medical treatment or support, they were subjected to immense suffering by families and healers resulting from chaining and beating (de-Graft Aikins et al. 2010). In the Magu district of Tanzania, the killing and victimisation of older women accused of witchcraft has been a problem for many years (Kowal and Suzman 2003). Another common social myth is that women deserve a lower social status than men (Economic Commission for Africa 2007). Despite international commitments to equality, women continue to occupy a lower social status, particularly in African society. At an individual level, older women often have less education, training, and resources than men. In these contexts, women are more likely to experience barriers in accessing health care and other social services, which is detrimental to their health and well-being. Despite the vital roles that women play in society, in many countries there is a gap between policy and practice with regard to equal opportunities and rights. The development of policies and programmes that support social inclusion is essential in ensuring that both men and women are able to realise their full potentials (Economic Commission for Africa 2007).

Another commonly held belief is that older people are unproductive, weak, helpless, and unable to absorb new information or learn new skills; they therefore have nothing to contribute and are an economic burden on society (Economic Commission for Africa 2007). Evidence to the contrary shows that instead of being an economic burden, older people are actually a source of wisdom, knowledge, skills, and experience, and are highly valuable to a country's economy (Lockwood 2003). In the future, it is projected that changes in the profile of the workforce and labour force shortage will result in the elderly constituting a larger share of the workforce (Pandey 2009; Robson 2001). Older people also contribute significantly in other work areas including the informal sector, agriculture, and as leaders, teachers, and volunteers in the community (Economic Commission for Africa 2007). A study in the Kagera district of Tanzania found that many older people were economically active beyond 75 years (Kowal and Suzman 2003). Almost 90% of respondents were still working on farms while 80% gained an income by selling their harvest (Kowal and Suzman 2003). Policies in developing countries should help support the economic potential of the elderly to enable those who are willing to work to stay economically active for longer. As steady gains are made in health, there has been an increase in longevity. In the absence of a secure income or sustainable livelihoods, financial insecurity is a key concern for older people. Investing in the health and well-being of the elderly will not only have long-term benefits for the economy, but will also help alleviate increasing poverty in old age.

Chronic diseases have a significant impact on the physical capabilities, social identities, and life trajectories of its sufferers (de-Graft Aikins et al. 2010). In Sub-Saharan Africa, studies on experiences of chronic illnesses show that these experiences are characterised by unhappiness, depression, spiritual distress, and suicidal tendencies (Joubert and Bradshaw 2006; Lloyd-Sherlock 2005; WHO 2002). Stigma has been shown to have very negative effects on health. In many African countries, physical chronic conditions such as diabetes, cancers, and epilepsy, and mental disorders (such as schizophrenia and psychosis) are heavily stigmatised. These psychological, emotional, and spiritual disruptions serve to undermine medical relations as well as illness management and self-care (de-Graft Aikins et al. 2010). Myths around the health of the elderly only serve to perpetuate these negative health outcomes. By providing health education to communities and caregivers, health systems can assist in challenging the myths that dominate about the elderly. In Africa, the ageing of the population is not a priority in national policies and programmes. In coming years population ageing will demand health programmes not only be re-orientated towards preventing, treating, and managing chronic diseases but also towards understanding the socio-cultural context in which older people are embedded, and its impact on the health care of the elderly.

3.6 Factors Affecting the Health of the Elderly

The health of a person is influenced by a variety of factors known as the social determinants of health. According to WHO (2008a), the social determinants of health are the "conditions in which people are born, grow, live, work, and age". These factors have been identified as shaping the health statuses of people over time. Some of the key determinants which have been identified as influencing health outcomes among individuals include: socio-economic conditions, education and employment opportunities, urban/rural residence, marital status, and social networks among others (refer to Fig. 3.1).

3.6.1 Socio-economic Conditions

In many countries, the elderly suffer high levels of poverty. Socio-economic conditions contribute significantly towards the health outcomes of older people. Poor socio-economic conditions in the form of material deprivation, combined with lack of quality medical health care, significantly increase people's chances of experiencing ill health in old age (Marmot 2005). However, due to the varying position of men and women in societies, the link between socio-economic status and health can only be understood in certain socio-cultural contexts. This is due to the fact that indicators of socio-economic status are heavily context-dependent, and certain aspects of male versus female status may have contradictory effects on health in different settings (Kuate-Defo 2006).

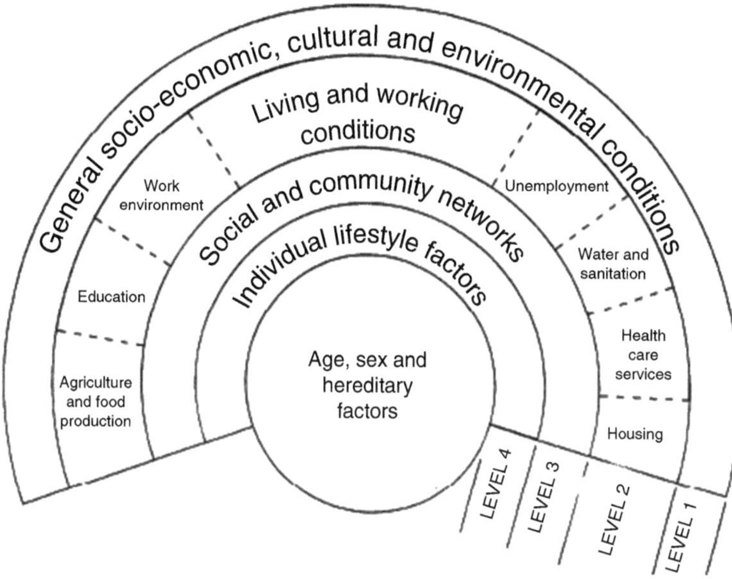

Fig. 3.1 The determinants of health and well-being. *Source*: WHO (2006)

Though indicators of socio-economic status tend to be context-dependent, numerous studies have highlighted the impact it has on the health and well-being of the elderly. This is no more apparent than in Africa, which is characterised by high levels of social inequality and widespread poverty. In Africa, the latest Human Development Report shows that a significant section of the population live on less than $1 a day (UNDP 2010). Consequently, research has found that men and women in lower- and middle-income countries report poorer health than those in higher-income countries (Nawi et al. 2010). In this context, poverty-related problems have been identified as contributing to a high level of morbidity (Lloyd-Sherlock 2005). Figure 3.2, which highlights the survival curve of elderly adults according to socio-economic status, shows that, although there is no significant difference in mortality among the different quintiles, the poorest group still has the lowest survival probability of all the groups. Several studies highlight that a positive relationship exists between socio-economic status and older people's health, well-being, and overall quality of life (Nawi et al. 2010; Mwanyangala et al. 2010). In addition to physical health, socio-economic status has also been linked to mental health including stress, depression, and anxiety. This was found among Black women in rural South Africa who are among the poorest in the country (Doyal and Hoffman 2009). In this context, low socio-economic status together with lack of education and unemployment were highlighted as important factors contributing to depression (Doyal and Hoffman 2009).

Socio-economic status is also cited as an important determinant of health-seeking behaviour among the elderly. A household survey in Nigeria found that the elderly

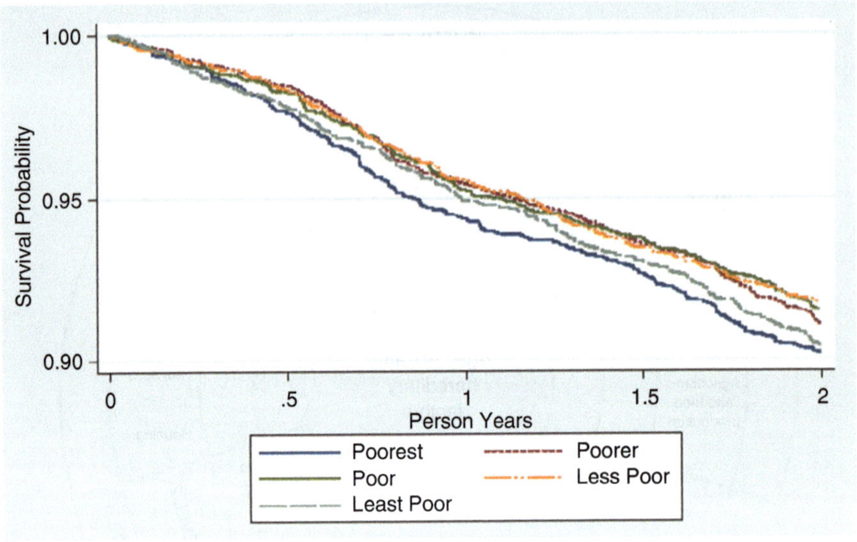

Fig. 3.2 Kaplan–Meier survival curve of elderly adults according to socio-economic status. *Source*: Khagayi et al. (2008)

from poorer households accessed less public and private health care than the elderly from wealthier households. Those with a low socio-economic status relied on home remedies from the family since they could not afford orthodox health care service (Abdulraheem 2007). Studies show that the incidence of poverty among the older population in Africa remains disproportionately high compared to other age groups; consequently, they tend to exhibit higher levels of mortality and morbidity (Economic Commission for Africa 2007). One of the reasons for the poor socio-economic conditions among the elderly is the breakdown of traditional support systems in the absence of strong public security mechanisms. In less developed countries, traditionally the composition of households was characterised by extended families. With rapid urbanisation, modernisation, and migration, extended households are becoming less common. Studies show that living with or near family is vital for the material, physical, emotional, and psychological support of elderly people (Zimmer and Dayton 2005). Changing household structures therefore impact significantly on their health and well-being.

Increasing longevity will have a substantial health, economic and social impact particularly on low- and middle-income countries, which are already burdened by the need to manage chronic care and non-communicable diseases among the older population (Nawi et al. 2010). Understanding the socio-economic determinants of health in Sub-Saharan Africa will not only be significant in reducing the impact it has on the health and well-being of the population, but may also form the basis for policies and plans aimed at reducing social inequities within societies.

3.6.2 Urban–Rural Residence

Place of residence has, for decades, been implicated as a determinant of health status (Zimmer et al. 2007). Despite this, very little research has examined urban–rural mortality and morbidity trends particularly in developing countries. As such, differences in health among the rural, semirural, and urban areas of Africa may not be so obvious (Kuate-Defo 2006). Research that does exist suggests that public health factors such as access to safe water, sanitation, and services make living in urban areas more beneficial in developing countries (UNICEF 2010; Cohen and Menken 2006). Those living in urban areas have also been shown to have a higher level of education, better job opportunities, and earn higher incomes (UNICEF 2010; Zimmer et al. 2007). These factors are considered to be robust predictors of health (Zimmer et al. 2007).

Africa is one of the least developed regions in the world. Existing data suggests that approximately two-thirds of people in Sub-Saharan Africa live in rural areas, relying primarily on subsistence agriculture and pastoralism for their livelihoods (Cohen and Menken 2006). In these contexts, chronic poverty becomes a risk factor for the health and well-being of people. The great majority of older people in Africa live in rural areas which are characterised by higher unemployment, few opportunities, and poor social protection (UNICEF 2010; Lloyd-Sherlock 2005). As a result, they have a higher level of disease prevalence than those in urban areas. A survey in rural KwaZulu-Natal found that older people living in rural areas experienced greater health problems (Hosegood and Timaeus 2006). In this context, non-communicable diseases accounted for 76% and 71% of deaths of women and men, respectively. Research shows that, in addition to variations in population health, life expectancy, and adult mortality, the average rate of self-reporting of good health and functional status is usually lower in rural than in urban areas (Kuate-Defo 2006). Rural residents are particularly vulnerable to poor health due to poor social infrastructure including inadequate and limited access to health care. In cases where health services are available, the health staff are over committed, staff training may be limited, and chronic disease management programmes may not be fully developed (Gomez-Olive et al. 2010). For older people this has been exacerbated by the breakdown of traditional social support systems. In many rural regions of Africa, water is a scarce commodity. Of those who lack access to improved drinking water sources, 84% live in rural areas (UNICEF 2010). There are also continuing disparities in sanitation between urban and rural areas. Across developing regions, sanitation coverage is 70% higher in urban areas than rural areas (UNICEF 2010). In these contexts access to water and sanitation are significant determinants of death and disease.

In recent years, population ageing has been accompanied by rapid urbanisation in Africa amidst deteriorating economic conditions (WHO 2008b). This has resulted in an increase in the number of informal or slum settlements. It is estimated that more than 70% of urban residents in Sub-Saharan Africa live in slums or slum-like conditions (Kyobutungi et al. 2010). The informal nature of these settlements implies that they are under-served by the public sector in the provision of basic services including health. Studies from various countries in Africa have shown that slum residents have poorer health outcomes than those in rural areas (WHO 2008b). A case study on the health and well-being of older people in Nairobi's urban slums found that desperate living conditions and lack of livelihood opportunities predisposed many to risky health-related behaviours such as smoking, alcohol consumption, and other substance abuse (Kyobutungi et al. 2010). Older people in this context are faced with additional challenges including weak social networks, loss of respect, and neglect. These factors, compounded with a lack of access to health services, predispose individuals to negative health outcomes. The unparalleled pace of urban growth has in recent years resulted in a vanishing rural–urban gap. In the context of Africa, Nigeria, Kenya, Tanzania, and Uganda are prime examples of countries that are characterised by vanishing rural–urban gaps and a growing urban population (WHO 2008b). Not only has this increased competition for resources, but it has also led to changes in patterns of marriage, fertility, and household structures, which perpetuates poverty. Several studies have shown that the population of urban poor is increasing rapidly (Takyi 2011). In countries such as Niger, Zambia, Kenya, and Senegal poverty rates for urban residents are almost comparable to those for rural residents (Takyi 2011). Emerging evidence suggests that higher risks for lifestyle-related diseases, such as diabetes and cardiovascular disease, are also beginning to arise in some urban areas in the developing world (Zimmer et al. 2007). It is projected that over half the population in Sub-Saharan Africa will live in urban areas by 2030 (Kyobutungi et al. 2010). This will give rise to various social and economic problems. The implications for the health of the elderly population are therefore significant.

3.6.3 Gender

Gender is one of the most significant socio-cultural factors influencing health and health-related behaviour (Evans et al. 2011). An extensive body of literature shows that in most countries women on average have longer life expectancies than men (Nawi et al. 2010). Nonetheless, despite their increased longevity they report experiencing poorer health outcomes than men (Nawi et al. 2010). In general, findings from North America, Europe, Asia, and several African countries show that older men report better health than older women (Nawi et al. 2010). However, gender and notions of masculinity play a huge role in men reporting better health status than women (Evans et al. 2011). For instance, men subscribe to a masculine role of being strong and autonomous, and these traditional notions of masculinity contribute to men's experience of their own health and health issues (Holroyd et al. 2005). These notions of masculinity intersect with other social determinants of health thus creating health disparities among men and women (Evans et al. 2011). Another common reason for these health disparities are the gender patterns in society, which affect the degree to which men and women are exposed to various risk factors, which impact their health in the long term (WHO 2003b). For instance, gender roles and norms mean that older women are less educated than men, have less formal work experience, and have less access to sources of income (Joubert and Bradshaw 2006). In every country, except for Botswana and Lesotho, illiteracy rates were found to be higher for females than males (Economic Commission for Africa 2007). Women are therefore less likely to participate in the formal labour market, and more likely to have a lower socio-economic status which has serious poverty and health implications for later life. A study examining self-reported health and physical functional status among older men and women living in 75 urban and rural localities of Cameroon found that older men reported better health than older women with the differences explained by the health advantage of their labour force and economic activity status (Kuate-Defo 2006). Similarly, older women reported being in poor health with functional limitations more than men, with the female disadvantage entirely explained by differences in their socio-economic status relative to older men (Kuate-Defo 2006).

The AIDS epidemic in Sub-Saharan Africa has resulted in changing household structures which have had significant implications for the roles and responsibilities of older women. In Nairobi and Zimbabwe older women are having to play the role of caregivers to orphaned grandchildren while providing economic support through their social pensions and old-age grants (Kyobutungi et al. 2010; Gomez-Olive et al. 2010; WHO 2002). Households headed by older women are therefore more likely to be impoverished and are twice as likely as other households to be caring for orphaned and vulnerable children (Economic Commission for Africa 2007). In some African countries, women often have no inheritance rights, which lead to poverty and isolation. Women are also more likely to experience discrimination in accessing health services and social security (Razzaque et al. 2010). As a consequence of these cumulative disadvantages, older women are more likely to report poorer health, disease, and disability than older men (Razzaque et al. 2010; Joubert and Bradshaw 2006). Health inequality studies between men and women in Africa and Asia, found that men had higher health scores than women across all age groups (Nawi et al. 2010). Socio-economic and demographic characteristics as well as gender discrimination were identified as key determinants of discrepancies in health across contrasting African and Asian settings. Similarly, in rural South Africa, women were significantly more likely to report lower health status as well as poor functionality than were men (Gomez-Olive et al. 2010). Though the basic diseases which afflict older men and women are the same, the rates, trends and types of diseases differ significantly (WHO 2003b). Mental health, musculoskeletal problems, and sensory impairments are found to be more common among women while cardiovascular diseases and cancers are more prevalent among older males (WHO 2003b).

According to Nawi et al. (2010) the differences between men and women are expected to widen over the next 30–40 years, with women constituting the majority of the older population. Ageing projections for 2050 show that there are higher proportions of women than men at each age band (60+, 65+, and 80+) (Economic Commission for Africa 2007). Given that women, particularly in the developing world, are greatly reliant on their husbands for economic and social support, this will mean that a large percentage of women will be at increased risk for poverty, neglect, and isolation (Nawi et al. 2010; Joubert and Bradshaw 2006). Elderly females will therefore bear the greater share of constraints caused by a lack of policies and programmes aimed at improving the health and well-being of the population (Kalasa 2001). This calls for a focus on the socio-economic situation of women and policies and plans to improve their well-being.

3.6.4 Education

The relationship between education and health has been well documented in many countries around the world. Cutler and Lleras-Muney (2007) suggest three broad explanations for the association between health and education. The first possible explanation is that poor health leads to lower levels of schooling, since poor health in childhood is linked to poor health in adulthood. Secondly, additional factors, such as family background or individual differences, both increase schooling and improve health. The third potential explanation for the link between education and health is that increased education directly improves health. However, Cutler and Lleras-Muney (2007) assert that evidence related to these explanations is mixed and it is unlikely that they fully explain the relationship between adult health and completed education.

Studies suggest that individuals with a higher level of education have lower self-reported morbidity rates for acute and chronic diseases (Cutler and Lleras-Muney 2007; Kuate-Defo 2006). One of the reasons for this is that higher education is associated with greater awareness of health problems and higher utilisation of health services. A study of immigrant and non-immigrant elderly people in America found that those with higher education levels experienced better physical and health outcomes, had increased health service utilisation, and health insurance coverage when compared with those with a lower level of education (Lum and Vanderaa 2010). Similarly, a study on health inequalities in Africa and Asia found that both men and women with higher levels of education consistently had higher health scores compared to those with lower levels of education (Nawi et al. 2010). In Nigeria, elderly people with higher education were more likely to exhibit health-seeking behaviour (Abdulraheem 2007). This is also linked to the fact that education is a key determinant of socio-economic status in later life. People with higher levels of education are more likely to participate in the formal labour market and earn higher wages; consequently, they are more likely to afford quality medical health care. In Cameroon, higher numbers of older respondents with no education or who were unemployed, reported poor health, functional limitations, or both, when compared to their educated and employed counterparts (Kuate-Defo 2006). Cutler and Lleras-Muney (2007) highlight that a significant relationship exists between mortality and years of schooling, and between self-reported fair and poor health status and years of schooling. For instance, better educated individuals are less likely to self-report a past diagnosis of chronic disease, are less likely to die from the most common chronic diseases, and are less likely to report anxiety or depression (Cutler and Lleras-Muney 2007). Individuals with an additional 4 years of education are less likely to experience negative health outcomes including heart disease and diabetes as well as negative health behaviours such as smoking, drinking, obesity, and drug use (Cutler and Lleras-Muney 2007).

Apart from formal schooling and tertiary education, research findings also draw attention to the importance of health education in creating awareness about the behaviours linked with poor health outcomes (Economic Commission for Africa 2007). Health education can promote healthy living which can reduce the risk of disease at older ages, while also assisting in improving physical well-being. The effects of education on health appear to be the same for men and women (Cutler and Lleras-Muney 2007). However, though the majority of countries around the world have achieved gender equality in education, females continue to face discrimination in many regions. These disparities have been shown to be slightly more marked in rural areas and among the poorest households (UNICEF 2010). Across Africa women have a lower literacy rate (84%) than men (90%) (United Nations 2010a). Income, which is related to education and employment, also continues to differ somewhat between men and women. There is still a substantial wage gap between employed men and women in most regions, with women earning less than men (United Nations 2010b). This reduces their overall social and economic power. As a consequence of these disadvantages, women are more likely to report poorer health and disease in later life than men (Razzaque et al. 2010).

3.6.5 Marital Status

The impact of marriage on health has been the focus of much attention over the past decade. A review of the literature suggests that marriage is beneficial to the health and well-being of individuals (Wood et al. 2007). People who are married also exhibit increased longevity and reduced risk of chronic illness in later life (Staton 2008). Marriage is believed to protect individuals from poor health outcomes in a variety of ways. Marriage offers economic advantages by allowing the combining of two incomes which improves economic well-being. This can contribute to greater access to health care and increased health care use (Wood et al. 2007; Beckett and Elliot 2002). The spouse may in some instance, play a role in monitoring and encouraging healthy behaviours, such as adhering to medical regimens, while discouraging unhealthy behaviours such as excessive drinking or substance abuse (Beckett and Elliot 2002). Men, in particular, physically benefit from the status of being married including an improvement in health status, increase in positive behaviours, and reduced negative physical symptoms (Staton 2008). Marriage may also be emotionally fulfilling and satisfy the need for social connection, integration, and regulation which may have physical and mental health implications (Wood et al. 2007; Beckett and Elliot 2002). In some societies where single people face stigma, marriage represents a way of adhering to cultural norms. This leads to social acceptance which has been associated with mental health benefits (Wood et al. 2007).

Men and women that are married are also more likely to have children. In Africa, children and grandchildren are often the main sources of support for older men and women. The wealth flows theory, developed by John Caldwell, is relevant in this context as survey data shows that fertility decisions in many parts of Africa are linked to parents perceived upward wealth flow in high fertility contexts (Kaplan and Bock 2001). This theory postulates that children add positively to the wealth, security, social, and political well-being of parents in old age (Kaplan and Bock 2001). The economically rational decision would therefore be to have as many surviving children as possible as this contributes to familial wealth flows. A study on the interactions between socio-economic status and living arrangements in predicting gender-specific health status among the elderly in Cameroon found that larger family size was associated with a protective effect on the self-rated health of older people, particularly men (Kuate-Defo 2006). In the absence of social security schemes, family becomes the main source of security for the elderly. In Africa, family therefore remains one of the most important institutions (Takyi 2011).

Many studies in Africa highlight the impact that marital status has on the health and well-being of individuals. Individuals that are married exhibit lower levels of mortality, morbidity, and mental health disorders than those that are never married, separated, divorced, or widowed (Beckett and Elliot 2002). A study on health inequalities between older men and women in Africa and Asia found that in all sites men and women who were not in current partnerships had lower health scores than those with partners (Nawi et al. 2010). Similarly, in rural South Africa, older people who were not married reported poorer health outcomes than those that were married (Gomez-Olive et al. 2010). In Cameroon it was found that single, widowed, or divorced elderly people are three times more likely to report poor health and functional limitations than those in monogamous marriages (Kuate-Defo 2006). A study on quality of life among older adults in rural Tanzania shows that poor health status, poor quality of life, and well-being were significantly associated with marital status (Mwanyangala et al. 2010). A person's marital status at age 48 is therefore a strong predictor of their chances of surviving to age 65 (Staton 2008). While married men have a 90% chance of surviving to age 65, divorced men have only a 65% chance of survival. Similarly, a married woman has a 95% chance of surviving to age 65, compared to an 80% chance of a never married woman (Staton 2008).

3.6.6 *Social Networks*

Social networks have a significant impact on the health of individuals. According to this perspective, the health of people is interdependent, and social networks influence health through a variety of mechanisms including the provision of social support, social engagement, and access to resources (Smith and Christakis 2008). However, research suggests that the causal relationship between social networks and health is complex involving behavioural, psychological, and physiological pathways (Gallegos-Carrillo et al. 2009). For instance, numerous studies have demonstrated that the connection between social relationships and health varies over the life course (Gallegos-Carrillo et al. 2009). Hence, as the population becomes demographically older and social networks change, new relationships may form which is likely to impact physical health and well-being. Individuals who lack social connections tend to suffer higher rates of morbidity and mortality, depression, and cognitive decline (Cornwell and Waite 2009). Previous research has also identified a wide range of indicators of social isolation that pose health risks, including living alone, feelings of loneliness, and infrequent participation in social activities (Cornwell and Waite 2009). In Botswana, older people who were living in small households and had limited social networks were found to have diminished cognitive and physical function, as well as increased risk of rapid deterioration and mortality (Clausen et al. 2007). Social networks can help mitigate the negative effects of loneliness and isolation that can be detrimental to health, particularly in old age. The presence of social networks has been shown to have a positive influence on cognition and to be protective against the development of dementia among older people (Crooks et al. 2009). Social networks also play an important part in the management of long-term or chronic conditions by influencing access to and utilisation of medical services (Vassilev et al. 2010). Among the elderly in Nairobi, social ties had an impact on both men and women accessing services. Survey data in this context showed that the proportion of older people seeking health services was relatively higher for older people who lived with at least one other adult, compared to those who lived alone (Mudege and Ezeh 2009). Social ties are particularly important in socially and economically deprived contexts where chronic conditions including AIDS have shifted the emphasis away from self-care and towards community and network-centred approaches. Policy makers should aim to strengthen formal care mechanisms and encourage informal care of the elderly population, particularly those with limited social support, in order to prevent detrimental health outcomes and premature mortality (Clausen et al. 2007).

Social determinants play a key role in influencing the health outcomes of the elderly. Socio-economic factors, urban–rural residence, gender, education, marital status, and social networks have been identified as shaping the health status of people over time (Chapman 2010). Understanding these determinants of poor health are therefore essential to addressing the health needs of ageing populations and improving their health trajectories (Nawi et al. 2010).

3.7 Access to Health Care

As the population grows older the demand for health services also increases. In Africa, data on geriatric service provision and utilisation is lacking (Joubert and Bradshaw 2006). Improving access to health care services for the elderly is therefore a major challenge. With population ageing, the disease burden has shifted to non-communicable diseases, increased levels of chronic diseases, growing levels of disability, and an increasing loss of physical and cognitive functioning (Suzman 2010).

This has resulted in increasing recognition that older people require services that are sensitive to their unique needs, and a move for health services to be re-orientated towards meeting the needs of the elderly. Until recently, most public health programmes in developing countries have been directed towards improving child health. Treatment and management of chronic illness and access to health care for the elderly have not been regarded as policy priorities, with health care systems spending only a small fraction of their budget on treating illness among older adults (Mwanyangala et al. 2010). This inaction by policy makers has led to inequalities in health care for older people.

At the individual level, the elderly face numerous barriers in accessing health care. Some of the barriers which have been identified include interpersonal relations and communication problems between health providers and elderly patients, and a lack of knowledge about services and treatment. A study on health-seeking behaviour in Kenya found that the negative attitudes of health care workers were associated with older people delaying seeking health care (Waweru et al. 2003). In Tanzania, 40% of older people reported that the tone of language used by medical staff was disrespectful and mocking, while over a third had to wait between 4 and 6 hours in order to see a doctor (HelpAge International 2008). In South Africa, older people expressed dissatisfaction with the quality of health care at the primary level, including inefficient appointment systems, long waiting times, and apparent lack of interest of staff in the health problems of the elderly (Joubert and Bradshaw 2006). Older people in urban and rural areas revealed that the quality of public health care services they received was of major concern including: shortage and unavailability of medication, unavailability of assistive devices, and perceived lack of respect, and sharing of information by the health personnel who attended to them (Joubert and Bradshaw 2006). In Kenya, 62% of older people reported buying over the counter drugs (Waweru et al. 2003). This high level of older people accessing over the counter drugs is indicative of the inefficiency of health services in meeting the needs of the elderly in developing countries. These constraints in health service provisions are exacerbated by the shortage of staff trained in the care and treatment of older people. In Africa, only a small proportion of health workers have specialist training in management of chronic illness, while knowledge of chronic illness among health workers in general is poor (de-Graft Aikins et al. 2010). A study on the perceptions and attitudes of medical students towards older patients in Tanzania found that 45% of respondents regarded older people as dependents, unpleasant, unhealthy, dull, and ugly (Kowal and Suzman 2003). It was noted that only 2% of these respondents had attended courses related to ageing, all of which were outside the country. This study concluded that a lack of geriatric teaching and exposure to geriatric medicine contributes to negative perceptions around the elderly and reduces the quality of services delivered.

Most health systems also lack the human capacity for chronic disease care. WHO lists Sub-Saharan Africa as one of the geographic areas least served by health care providers (Livesay 2007). In 15 countries in Sub-Saharan Africa, there are five or fewer physicians per 100,000 people, and in 17 countries, there are 50 or fewer nurses per 100,000 people (Global Health Council 2009). This is below the

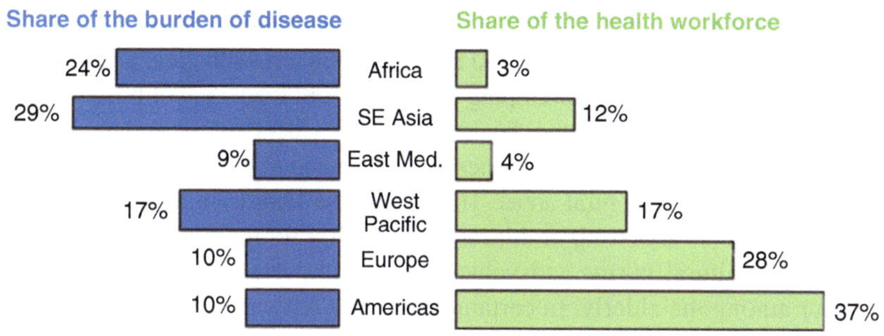

Fig. 3.3 Global disease burden and health care workforce. *Source*: Global Health Council (2009)

recommended 20 physicians per 100,000 and 100 nurses per 100,000 which significantly impacts on effective health care provision. Figure 3.3 shows that, while Africa is home to 24% of the global burden of disease, it only has 3% of the total global health workforce. In contrast, America has close to 40% of the total global health workforce but only accounts for 10% of the global burden of disease. According to the Global Health Council (2009) in Africa alone there is a shortage of 1.5 million health care workers. Nurses, in particular, play a critical role in geriatric care. With the projected increase in the elderly population, health systems in African countries will require greater mobilisation of the workforce.

Economic barriers to accessing services and treatment are often experienced by older people who lack financial and social support. In Africa, the economic situation of the elderly is closely tied with the overall situation of the extended family. With a reduced ability to generate resources, changing family structures, unfavourable social and economic conditions, and the AIDS epidemic, the elderly are becoming increasingly more vulnerable in their old age. The transportation costs in accessing services, the cost of treatment and services, and the overall affordability of health care can therefore become a barrier to health. In Kenya, 73% of older people reported lack of money as hindering their access to health care (Waweru et al. 2003). In countries that offer free health care, lack of knowledge about these services prevents older people from accessing them and exacerbates the economic costs of seeking services. In Tanzania, where health services for the elderly were free, 94% of older people were charged during their hospital visit, and 30% were unaware of how to apply for free health (HelpAge International 2008). Almost half the respondents reported having to pay their own transport fees to get to the hospital. Though a few select countries in Africa provide a state pension or government grants, many older people use this to support their families, particularly in households afflicted by AIDS. Consequently, this form of income is often insufficient to support the health needs of the elderly. Lloyd-Sherlock (2005) therefore argue that there is a need to mainstream older people into discussions about health sector reform and the composition of selective health care packages and essential drug programmes.

In rural areas geographical isolation, lack of public transportation, and poor social infrastructure, including inadequate and limited access to health care, all contribute to negative health outcomes. In developing countries more than 75% of physicians, 60% of nurses, and nearly 60% of other health service providers are located in urban areas (Global Health Council 2009). Countries in Africa need to prioritise the expansion of services to populations in rural areas. This will in the long term address some of the economic barriers to accessing health care, particularly for the frail elderly.

In Africa, cultural norms and values play an important role in health-seeking behaviour among the elderly. In certain traditional African households women are only allowed to visit health facilities with the consent of their spouse or head of the household (Razzaque et al. 2010). In these settings women disproportionately bear the burden of health inequalities and are presented with significant barriers to accessing medical care. A range of medical systems provide chronic disease care in many African countries, including biomedical services, traditional medicine, and faith healing systems (de-Graft Aikins et al. 2010). Based on their cultural beliefs, in these settings, older people may seek treatment from traditional healers or diviners, which acts as a barrier to accessing western medicine. Aside from cultural beliefs, medical services are often inaccessible to the poor, and poverty intensifies the use of traditional healers who offer services at a cheaper rate (de-Graft Aikins et al. 2010). In South Africa, it is estimated that 60–80% of the population consult a traditional healer before going to a primary health care practitioner (Truter 2007). Socio-cultural barriers including stigma and lack of knowledge about health conditions and services prevent many older people from accessing health care. By orientating health care services towards the unique needs of the elderly, countries can help ensure that the older population live healthy, productive lives. Governments need to recognise that access to health care that can address the health issues of the elderly is important.

3.8 What Do We Know About the Health of the Elderly?

In Africa, morbidity and mortality statistics are an important indicator of the health status of the population (Joubert and Bradshaw 2006). Health data is usually obtained from general sources for demographic data including vital registration systems, general surveys and censuses, as well as from specialised sources such as administrative health records, clinical trials, and epidemiological studies. However, adequate systems that produce statistics on a regular basis are grossly lacking (Rao et al. 2006). Very few countries in Africa have functioning vital statistics systems that produce usable information on mortality and morbidity (Velkoff and Kowal 2006). In addition, there is a dearth of qualitative and quantitative research on the health status of the older population. Gaining an accurate picture of the mortality profile of the elderly is, therefore, problematic. In this section, every effort will be made to present data on mortality and morbidity among the elderly in Africa. These will be highlighted for selected chronic conditions including cardiovascular disease, respiratory disease, and diabetes.

Table 3.1 Estimates of leading causes of death in Sub-Saharan Africa at age 60 years and older: comparison between males and females, 2000

	Cause of death in males (*N*=953,391)	Total deaths (%)	Cause of deaths in females (*N*=973,542)	Total deaths (%)
1	Ischemic heart disease	13.4	Cerebrovascular disease	17.0
2	Cerebrovascular disease	10.8	Ischemic heart disease	13.4
3	Lower respiratory infections	6.5	Lower respiratory infections	5.6
4	COPD	61	Nephritis and nephrosis	3.8
5	Prostate cancer	4.0	Diabetes mellitus	3.7
6	Tuberculosis	3.1	COPD	3.5
7	HIV/AIDS	2.9	Hypertension heart disease	3.2
8	Nephritis and nephrosis	2.9	Diarrheal diseases	2.6
9	Diarrheal diseases	2.3	HIV/AIDS	2.3
10	Diabetes mellitus	1.9	Cervix uteri cancer	2.3

Source: Rao et al. (2006)

3.8.1 Causes of Mortality and Morbidity Among the Elderly in Sub-Saharan Africa

Though there is dearth of information on the health of the elderly, what research does exist suggests that age-specific mortality rates from chronic diseases are higher in Africa than in all other regions of the world, for both men and women (de-Graft Aikins et al. 2010). Estimates of leading causes of death for those aged 60 years and older for 2000 show that, for both males and females, ischemic heart disease, cerebrovascular disease, and lower respiratory infections ranked in the top three leading causes of death (refer to Table 3.1). Over the next 10 years it is projected that the continent will experience the largest increase in death rates from cardiovascular disease, respiratory disease, and diabetes.

3.8.1.1 Cardiovascular Diseases

Cardiovascular diseases affect the heart and blood vessels in the body, and are usually characterised by the loss of elasticity in the arteries due to blockage or hardening. This family of diseases includes rheumatic heart disease, hypertensive disease, and cerebrovascular diseases. Cardiovascular disease is one of the most common causes of mortality and morbidity for older men and women worldwide (Okunola et al. 2011). In 2002, an estimated 29% of deaths across the world and 43% of global mortality and morbidity, calculated from disability-adjusted life years, were due to cardiovascular disease (Mbewu and Mbanya 2006). Research suggests that the majority of cardiovascular disease is occurring in developing country contexts (Okunola et al. 2011). A total of 78% of global mortality and 86% of mortality and

Table 3.2 Phases of the epidemiological transition in cardiovascular diseases

Phase of epidemiologic transition	Deaths from circulatory disease	Circulatory problems	Risk factors
Age of pestilince and famine	5–10	Rheumatic heart disease, infection, and deficiency-induced cardiomyopathy	Uncontrolled infection, deficiency conditions
Age of receding pandemics	10–35	As shown above, plus hypertensive heart disease and hemorrhagic stroke	High-salt diet leading to hypertension; increased smoking
Age of degenerative and man-made diseases	35–55	All forms of stroke; is chemic heart disease	Atherosclerosis from fatty diets, sedentary lifestyle; smoking
Age of delayed degenerative diseases	Under 50	Stroke and ischemic heart disease at older ages	Educational and behavioural changes leading to fewer risk factors

Source: Bradshaw et al. (2006)

morbidity as a result of cardiovascular disease was found in developing countries (Mbewu and Mbanya 2006). While the epidemiology of cardiovascular diseases in Africa may not be a true representation of the pattern of heart disease in the continent due to a lack of data, studies suggest that cardiovascular disorders are the second-most common cause of mortality after infectious diseases and a major cause of chronic illness and disability (de-Graft Aikins et al. 2010; Mocumbi and Ferreira 2010). In 2001, cardiovascular disease accounted for 9.2% of deaths in Africa, principally due to hypertension, stroke, cardiomyopathy, and rheumatic valve disease (Livesay 2007). In South Africa, ischemic heart disease was among the leading cause of death for both elderly men and women in the population (Joubert and Bradshaw 2006).

The process responsible for the shift in mortality from cardiovascular diseases has been termed the epidemiological transition (refer to Table 3.2). In the first phase, rheumatic heart disease, infection, and deficiency-induced cardiomyopathy are the main circulatory problems. During the age of receding pandemics, hypertensive heart disease and hemorrhagic stroke emerge. All forms of stroke and ischemic heart disease emerge during the age of degenerative heart disease, resulting from growing obesity and diabetes. In the age of delayed degenerative diseases, stroke and ischemic heart disease occur at older ages. Since most cardiovascular disease cases in Africa are reported to be rheumatic, hypertensive and deficiency-induced, many studies suggest that Africa is largely in the first phase of the transition (Gaziano et al. 2006). However, in contrast, evidence from East Africa shows that countries and regions in Africa may be at different stages of transition as reflected in their cardiovasular disease profile (Ogeng'o et al. 2011).

This concept highlights that three main drivers fuel this transition. Firstly, declining infant and child mortality leads to rapid demographic changes, resulting in large increases in the number of individuals surviving until middle and older age, when chronic diseases manifest. Secondly, death rates from communicable diseases fall as a result of improved primary health care services and socio-economic development. Lastly, socio-economic development results in changes in the environmental and behavioural determinants of cardiovascular disease. Longer life expectancy also leads to longer periods of exposure to these determinants, increasing the prevalence of cardiovascular disease. This concept is helpful in understanding the causes leading to the emergence of the cardiovascular disease burden in Africa by focusing on the socio-economic, environmental, and behavioural determinants of cardiovascular disease. In 2020, it is estimated that cardiovascular disease will be the leading cause of the global health burden, accounting for 73% of global mortality and 56% of total morbidity (Mbewu and Mbanya 2006). The elderly are expected to account for a large majority of the mortality and morbidity burden. African countries will need to intensify preventative strategies for cardiovascular disease.

3.8.1.2 Respiratory Diseases

Chronic respiratory diseases affect the airways and other structures of the lungs. Preventable chronic respiratory diseases include asthma and respiratory allergies, chronic obstructive pulmonary disease, occupational lung diseases, cancer, sleep apnoea syndrome, and pulmonary hypertension (WHO 2009). Though the actual burden in developing country contexts is difficult to assess due to a lack of routine data systems, WHO 2009 estimates that more than 50% of people who suffer from chronic respiratory diseases live in low- and middle-income countries. Respiratory diseases, particularly lower respiratory tract infections, can be the most severe and are leading causes of death among adults over the age of 60 and people with compromised immune systems (refer to Fig. 3.4) (Global Health Council 2008). For the age groups over 60, rates of death from lower respiratory disease more than double for each decade of life (Speizer et al. 2006). A study on the health status of the elderly in Nairobi, Kenya, found that, among 400 non-institutionalised elderly persons, 68% suffered from respiratory conditions (Waweru et al. 2003).

Aside from lower respiratory tract infections, chronic obstructive pulmonary disease dominates all other chronic respiratory diseases (Speizer et al. 2006). Worldwide, it accounts for more than 10% of lost disability-adjusted life years. In 2005, there were more than three million deaths caused by chronic obstructive pulmonary disease (van Gemert et al. 2011). Ninety percent of these occurred in low- and middle-income countries. Mortality from chronic obstructive pulmonary disease is low before age 45; however, its incidence increases dramatically with age. Over age 45, death rates increase from 50 to 200 per 10,000 individuals and are consistent across age groups in men and women. In South Africa, chronic obstructive pulmonary disease is among the top three leading causes of death for elderly men while accounting for a large percentage of the mortality burden among women (Joubert and Bradshaw 2006).

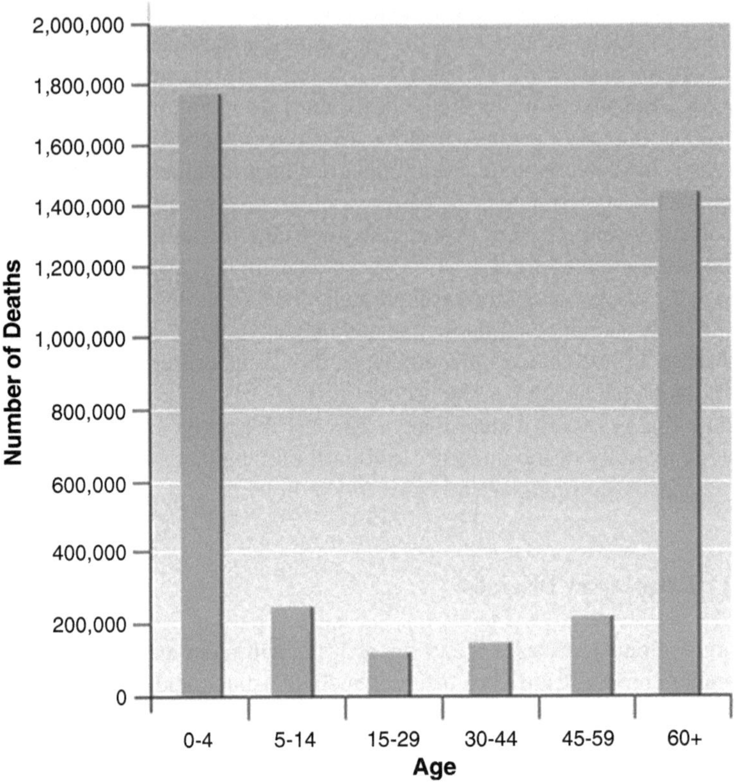

Fig. 3.4 Global deaths due to lower respiratory infections by age, 2008. *Source*: Global Health Council 2008

Over the next few years, mortality and morbidity from chronic respiratory disease is expected to increase. Though data on the incidence of chronic respiratory disease is scarce in Africa, many of the respiratory conditions such as chronic obstructive pulmonary disease are preventable. Policy and research needs to be directed towards implementing effective measures that ensure early detection and prevention of these conditions.

3.8.1.3 Diabetes Mellitus

Diabetes mellitus is one of the most common non-communicable diseases. The two main types of diabetes are Type 1 diabetes which occurs if the body's immune system kills the cells that produce insulin in the pancreas, and Type 2 diabetes which occurs when the pancreas fails to produce enough adequately functioning insulin to enable the glucose from food to enter the body cells and be used as a source of energy (Motala and Ramaiya 2010). Worldwide, close to 300 million people, constituting 6.6% of the adult population has diabetes. It is projected that this figure will rise by 54% by the year

Table 3.3 Global burden of diabetes prevalence, 2010–2030

	2010	2030
Total world population (billions)	7.0	8.4
Adult population (20–79 years, billions)	4.3	5.6
Diabetes (20–79 years)		
Global prevalence (%)	6.6	7.8
Comparative prevalence (%)	6.4	7.7
Number of people with diabetes (millions)	300	438

Source: Diabetes Atlas (2010)

Table 3.4 Diabetes prevalence in the African region, 2007–2025

	2007	2025
Total world population (millions)	747	1,088
Adult population (20–79 years, millions)	336	537
Diabetes (20–79 years)		
Regional prevalence (%)	3.1	3.5
Comparative prevalence (%)	3.6	4.5
Number of people with diabetes (millions)	10.4	18.7

Source: Diabetes Atlas (2007)

2030, with 438 million people having diabetes (refer to Table 3.3). The disease has been identified as the fourth or fifth leading cause of death in most high-income countries, while evidence suggests the effect of diabetes is particularly severe in low- and middle-income countries (Diabetes Atlas 2010; Mbanya and Ramiaya 2006).

Diabetes contributes significantly to the mortality and morbidity burden in Africa. Mortality rates for diagnosed diabetes in Africa vary greatly from 7.6 to 41% (Motala and Ramaiya 2010). However, studies suggest that Africa has the highest age-specific mortality rate in the world. With regard to morbidity, it is estimated that the complications of diabetes lead to: 4.51 million people with eye complications, 907,500 with cardiovascular disease, 399,300 with cerebrovascular disease, 423,500 with blindness, and 2.23 million with kidney damage (Motala and Ramaiya 2010). The complications of diabetes account for a large proportion of the mortality and morbidity burden among the elderly in Africa. A profile of elderly patients with diabetes in Nigeria found that diabetic foot disease (30.3%) was one of the main reasons for admission in the elderly group (Ikem et al. 2010). Diabetic foot disease (36.8%) accounted for the major cause of death in the elderly group (Ikem et al. 2010).

Though the availability of prevalence data is limited, age-specific levels of diabetes and hypertension in Sub-Saharan Africa have been shown to be higher than in most Western European countries (Mbanya and Ramiaya 2006). With data derived from South Africa, Tanzania, Ghana, Cameroon, and Sudan, the Diabetes Atlas (2007) estimated that 10.4 million people, constituting 3.1% of the adult population, had diabetes in the African Region in 2007 (refer to Table 3.4). In 2010, over 12 million people were estimated to have diabetes and 330,000 people were expected to die from diabetes-related conditions in Sub-Saharan Africa (Motala and Ramaiya 2010). The number of people with diabetes is projected to almost double in the next 20 years reaching 23.9 million by 2030.

Studies suggest that people aged 45–59 are 8.5 times more likely to develop diabetes than those aged 15–29, while people aged 60 and over are 12.5 times more likely to develop the disease (Motala and Ramaiya 2010). At a global level, it is projected that, by 2030, there will be 188 million people with diabetes aged 40–59, and 196 million people with diabetes in the 60–79 age group (Diabetes Atlas 2010). More than 80% of these people will be found in developing countries. In Sub-Saharan Africa the highest prevalence of diabetes is found among people of Indian origin, followed by native Africans (Motala and Ramaiya 2010). The prevalence of complications related to diabetes, however, does not vary by ethnicity. Untreated diabetes leads to various long-term complications, including blindness, kidney disease, neural vascular damage, coronary artery, and peripheral vascular disease, leading to amputations, stroke, and heart attacks (Motala and Ramaiya 2010). These complications result in increasing disability and reduced life expectancy among the elderly.

Cardiovascular disease, respiratory disease, and diabetes contribute significantly towards the mortality and morbidity burden of the elderly in Africa. With population ageing, African health systems have to be re-structured to address the burden of chronic illness. Current barriers to the health care system include overburdened health-care systems that are unable to meet the needs of chronic, non-communicable diseases, lack of resources, poor infrastructure, lack of donor attention, and poor governance. The availability of highly active anti-retroviral therapy has now also meant that AIDS is becoming a chronic disease (Gomez-Olive et al. 2010). A person diagnosed as HIV-positive at an early age can therefore expect to live well into their sixties. Research suggests that poorly managed HIV infection exacerbates ageing diseases and contributes to the development of opportunistic infections (including viral, bacterial, and fungal infections) leading to increased mortality (Mills et al. 2011). If non-communicable diseases are not addressed soon they, will over the next few years, overwhelm the fragile health infrastructure in Africa.

3.9 Risk Factors for Non-communicable Diseases

Chronic diseases are caused by a multitude of factors. In Africa underlying socio-economic, cultural, political, and environmental determinants for chronic diseases include globalisation, rapidly increasing urbanisation, westernisation, and population ageing. These determinants contribute to a change in lifestyle, diet, and health-related behaviour. In many regions these factors are exacerbated by poverty and lack of government programmes for the prevention of chronic diseases. These underlying determinants contribute to common modifiable risk factors for non-communicable diseases, including cardiovascular disease, diabetes, and respiratory disease (Fig. 3.5).

At the primary level, factors related to lifestyle and consumption patterns have been cited as important determinants of ill health (Lloyd-Sherlock 2005).

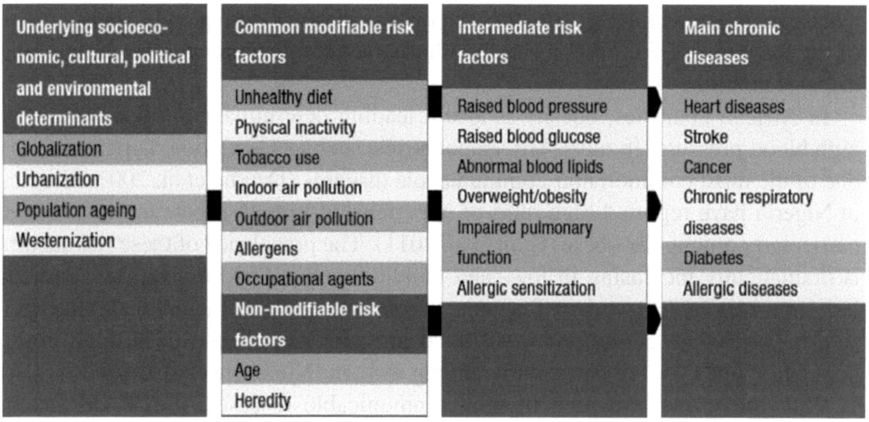

Fig. 3.5 Causes of chronic diseases. *Source*: WHO (2001)

WHO (2004) highlights that lifestyle and behaviour are primary determinants of chronic illness and have the potential to prevent, initiate, or advance these conditions and their associated complications. Physical inactivity and unhealthy diets that are high in saturated fats, sugars, and animal products, and low in staple foods including cereals, fruits, and vegetables are among the leading causes for non-communicable diseases (Lock et al. 2010). Worldwide, six risk factors related to nutrition, including high blood glucose and high blood pressure, account for 19% of deaths (Lock et al. 2010). Though nutrition patterns in Sub-Saharan Africa are influenced by many factors, including individual preference, culture, and price, the availability and accessibility of food are key factors that shape dietary patterns. A study in Cape Town, South Africa, found that a larger proportion of life spent in the city was associated with an increased consumption of fat—a risk factor for cardiovascular disease (Steyn and Damasceno 2006).

Physical activity is a key determinant in improving blood pressure and overall body weight, thereby reducing the risk of heart disease, diabetes, osteoporosis, and certain cancers. Though few studies on the physical activity patterns of people in Africa have been published, existing research suggests that the amount of physical exercise has been decreasing across the continent (Steyn and Damasceno 2006). Physical activity has been shown to have many health-promoting benefits and has a direct, independent role in decreasing the risk of cardiovascular disease mortality (Steyn and Damasceno 2006). Changing dietary patterns and an inactive or sedentary lifestyle have led to an increase in the prevalence obesity. Findings show that almost a quarter of elderly Africans are overweight while 56% of older South Africans are obese (Kimokoti and Hamer 2008). Despite its negative impact on health, in many Sub-Saharan Africa countries, an increased level of body fat is valued due to its association with beauty, prosperity, and dignity (Stern et al. 2010). In contrast, weight loss is perceived to be a sign of ill health or poverty. In recent years weight loss has been regarded as a source of stigma and a sign of disease due to its association with HIV and AIDS (Stern et al. 2010). Consequently obesity is one of the leading risk factors for chronic illness including cardiovascular disease and diabetes mellitus.

In African countries, obesity is also a leading determinant of hypertension or high blood pressure. In most African countries, such as Cameroon, hypertension is one of the most common non-communicable diseases (Dzudie et al. 2007). Studies in Nigeria have reported high rates of hypertension in both urban and rural areas, particularly among the obese (Ejim et al. 2011). The prevalence of these risk factors is dramatically increasing in low- and middle-income African countries, particularly in urban areas. In urban Tanzania, chronic disease risk factors including obesity, hypertension, cholesterol, and blood pressure were found to be high in the population aged 44–66, particularly among women (Njelekela et al. 2009).

With the growing burden of non-communicable diseases, the World Health Assembly endorsed a global strategy on diet, physical activity, and health in May 2004. The four main objectives of the strategy were to:

- Reduce the risk factors for non-communicable diseases resulting from unhealthy diets and physical inactivity through public health action, health promotion, and disease-preventing measures
- Ensure greater awareness and understanding of the influences of diet and physical activity on health, and of the positive impact of preventive interventions
- Encourage the development, strengthening, and implementation of global, regional, national, and community policies and action plans to improve diets and increase physical activities, and actively engage all sectors
- Monitor scientific data and key influences on diet and physical activity to support research in this domain to enhance and sustain health

While the global strategy on diet, physical activity, and health addressed diet and physical activity as the two main risks for non-communicable diseases, it also recognised the role of other health-promoting factors such as the avoidance of tobacco and control of alcohol consumption, among others. Research suggests that increased alcohol and tobacco use are critical risk factors for non-communicable diseases (Steyn and Damasceno 2006). Sensitivity to the effects of alcohol increases with age since older people have a decreased ability to develop tolerance to increasing amounts of alcohol and are more likely to achieve a higher blood alcohol concentration. Tobacco use has been shown to cause around 25 life-threatening diseases and is a major risk factor in 8 of the top 16 causes of death among older people (Economic Commission for Africa 2007).

In contrast to developed countries, studies in Africa reveal high rates of smoking and alcohol consumption among men and women aged 50 years and older in rural areas, which put them at risk of non-communicable diseases. A cross-sectional study in three rural sites in Malawi, Rwanda, and Tanzania found that smoking rates among older men and women were higher than among adults under 50 (Negin et al. 2011). A total of 21% of older women and 37% of older men were found to smoke, while alcohol consumption among older women was 45%. Examining a set of five risk factors, more men aged 50 and older (50%) had two or more risk factors than men under 50 (26%), while 52% of women aged 50 and older had two or more risk factors, compared with 24% of women under 50 (Negin et al. 2011). Similar findings were found among older people in South Africa. Among people aged 65 and over, 46% of men and 20% of women reported being current consumers of alcohol. Alcohol dependence was assessed at 23% and 12% for men and women, respectively. With regard to smoking, 35% of men and 7% of women reported smoking daily while over a quarter of all women and one-fifth of all men were exposed to tobacco smoke by sharing a household with smokers. These findings reiterate the fact that the health of older adults in Africa has been neglected. Policies and programmes need to recognise the importance of reaching out to older adults with relevant messages to encourage lifestyle changes in order to improve general health.

In the absence of policies and interventions, risk factors for chronic diseases will overwhelm the already burdened health systems in Africa. Determinants of the growing prevalence of cardiovascular disease, diabetes, and chronic respiratory disease in Africa are similar to those reported elsewhere, namely: family history of chronic illness, environmental factors, physical inactivity, increasing obesity, hypertension, diets high in cholesterol, high sugar and salt intake, tobacco use, and other unhealthy lifestyle and behavioural patterns (Mocumbi and Ferreira 2010). Mbewu (2009) argues that health promotion, including primary prevention focusing on these major risk factors, is needed at the community and primary health care level. By lowering high blood pressure, lowering high cholesterol, controlling blood glucose levels, preventing tobacco and alcohol use, controlling body weight, promoting a healthy lifestyle, and maintaining healthy eating behaviours, people at risk can avoid developing non-communicable diseases and their related complications.

3.10 Conclusion

More than a decade into the twenty-first century, there remains glaring disparities in health status and access to health care. The situation is particularly serious for the older population. For African countries, population ageing will exacerbate the challenges faced by health systems, including infrastructural limitations, shortages of appropriately trained health care workers, and a lack of policies and programmes directed at treating and managing non-communicable diseases. Unlike in developed countries, the principal health challenges in Africa have stemmed from the AIDS epidemic. With a growing elderly population, countries will now be faced with a cumulative burden of communicable and chronic diseases, along with injuries, disabilities, and mental health disorders associated with the ageing process. Governments will need to reallocate budgets towards non-communicable diseases to ensure that the gains made in health over the past decade are not eroded. Mobilisation of the health care workforce and training of workers in specialist chronic illness and geriatric care are essential in reorientating health care services towards the unique needs of the elderly. In addition, further research and epidemiological investigation on non-communicable diseases among the elderly is urgently needed. This is essential for the planning and implementation of prevention and control strategies for non-communicable diseases.

In addition to the considerable implications of population ageing on the health systems, in Africa, economic and social security for the elderly are policy concerns. Traditional caring and social support systems are increasingly facing strain. In the absence of social security systems, older people are faced with various economic challenges which impact their health and well-being. Older people are playing significant roles in providing emotional and economic support to children, grandchildren, and relatives in households afflicted by AIDS. In other regions, the elderly contribute towards the economy in other ways. Policy makers need to acknowledge that people are living for longer and can make significant contributions towards society. Prioritising the needs of the elderly is essential in ensuring they live healthy and productive lives.

Countries in Africa continue to face significant challenges in implementing health services that are accessible to the older population. While health systems are likely to be visited by increasing number of older people with chronic health problems, only a few countries have put into place plans or policies to deal with this chronic disease burden. South Africa, Tanzania, Cameroon, and Mauritius are among the few African countries that have responded to the call for action (de-Graft Aikins et al. 2010). For most countries in the African region, there is a gap between policy makers recognition of a national chronic disease burden, and the development and implementation of chronic disease policies (de-Graft Aikins et al. 2010). In the long term, sustainable progress will only be achieved with greater attention directed to capitalising on the links between health, ageing, and the recognition of fundamental human rights (Robinson et al. 2007). To effectively integrate the increasing number of older people in the development process within Africa will require collaboration and partnerships between the public sector, private sector, and civil society (Economic Commission for Africa 2007). It is imperative that countries recognise that the health of a population has far-reaching implications. As Robinson et al. (2007, p. xii) observed: "a country's health and its concern for its ageing population are fundamental not only to its wellness, but also to its social cohesion, its prosperity, and perhaps even to its political stability."

References

Abdulraheem, I. S. (2007). Health needs assessment and determinants of health-seeking behaviour among elderly Nigerians: a house-hold survey. *Annals of African Medicine, 6*(2), 58–63.

Aboderin, I. (2008). *Advancing health service provision for older persons and age-related non-communicable disease in Sub-Saharan Africa: Identifying key information and training needs.* Oxford: Oxford Institute of Ageing. http://www.instituteofageing.uct.ac.za/news/REPORT%20OF%20ABUJA%20DIALOGUE%20DRAFT.pdf. Accessed 7 Apr 2011.

Awofeso, N. (2010). Re-defining "Health". Commentary on editorial by Üstün and Jacob (2005; 83:802). *Bulletin of the World Health Organisation.* http://www.who.int/bulletin/bulletin_board/83/ustun11051/en/. Accessed 9 Apr 2011.

Beckett, M., & Elliot, M. N. (2002). Does the association between marital status and health vary by sex, race, and ethnicity? http://www.rand.org/pubs/drafts/2008/DRU2869.pdf. Accessed 10 Jul 2011.

Boerma, T. (2010). The INDEPTH WHO-SAGE collaboration-coming of age. *Global Health Action, 2*, 3–4.

Bradshaw, D., Schneider, M., Norman, R., & Bourne, D. (2006). Mortality patterns of chronic diseases of lifestyle in South Africa. http://www.mrc.ac.za/chronic/cdlchapter2.pdf. Accessed 10 Jul 2011.

Chapman, A. R. (2010). The social determinants of health, health equity, and human rights. *Human and Human Rights, 12*(2), 17–30.

Clausen, T., Wilson, A. O., Molebatsi, R. M., & Holmboe-Ottesen, G. (2007). Diminished mental- and physical function and lack of social support are associated with shorter survival in community dwelling older persons of Botswana. *BioMed Central Public Health, 7*, 144.

Cohen, B., & Menken, J. (Eds.). (2006). *Aging in Sub-Saharan Africa: Recommendations for furthering research*. Washington DC: The National Academies Press.

Cornwell, E. Y., & Waite, L. J. (2009). Social disconnectedness, perceived isolation, and health among older adults. *Journal of Health and Social Behavior, 50*(1), 31–48.

Crooks, V. C., Lubben, J., Petitti, D. B., Little, D., & Chiu, V. (2009). Social network, cognitive function, and dementia incidence among elderly women. *American Journal of Public Health, 98*(7), 1221–1227.

Curry, L. A., Nembhard, I. M., & Bradley, E. H. (2009). Qualitative and mixed methods provide unique contributions to outcomes research. *Circulation, 119*(10), 1442–1452.

Cutler, D. M., & Lleras-Muney, A. (2007). Education and health. National Poverty Center Policy Brief number 9. http://www.npc.umich.edu/publications/policy_briefs/brief9/policy_brief9.pdf. Accessed 18 Mar 2011.

de-Graft Aikins, A., Unwin, N., Agyemang, C., Allotey, P., Campbell, C., & Arhinful, D. (2010). Tackling Africa's chronic disease burden: from the local to the global. *Globalisation and Health, 6*, 5.

Diabetes Atlas. (2007). African region. International diabetes federation. http://da3.diabetesatlas.org/index747c.html. Accessed 15 Jun 2011.

Diabetes Atlas. (2010). Global burden. International diabetes federation. http://www.diabetesatlas.org/content/diabetes-and-impaired-glucose-tolerance. Accessed 15 Jun 2011.

Doyal, L., & Hoffman, M. (2009). The growing burden of chronic diseases among South African women. *Continuing Medical Education, 27*(10), 456–458.

Dzudie, A., Kengne, A. P., Mbahe, S., Menanga, A., Kenfack, M., & Kingue, S. (2007). Chronic heart failure, selected risk factors and co-morbidities among adults treated for hypertension in a cardiac referral hospital in Cameroon. *European Journal of Heart Failure, 10*(4), 367–372.

Economic Commission for Africa. (2007). The state of older people in Africa. African Center for Gender and Social Development. http://social.un.org/index/LinkClick.aspx?fileticket=F7VUCX8uUQ0%3D&tabid=375. Accessed 05 Mar 2011.

Ejim, E. C., Okafor, C. I., Emehel, A., Mbah, A. U., Onyia, U., Egwuonwu, T., Akabueze, J., & Onwubere, B. J. (2011). Prevalence of cardiovascular risk factors in the middle-aged and elderly population of a Nigerian rural community. *Journal of Tropical Medicine*. doi:10.1155/2011/308687.

Evans, J., Frank, B., Oliffe, J. L., & Gregory, D. (2011). Health, Illness, Men and Masculinities (HIMM): a theoretical framework for understanding men and their health. *Journal of Men's Health, 8*(1), 7–15.

Ferreira, M., & Makoni, S. (2002). Towards a cultural and linguistic construction of late-life dementia in an urban African population. In S. Makoni & K. Stroeken (Eds.), *Ageing in Africa. Sociolinguistic and anthropological approaches*. Aldershot: Ashgate.

Gallegos-Carrillo, K., Mudgal, J., Sanchez-Garcia, S., Wagner, F. A., Gallo, J. J., Salmeron, J., & Garcia-Pena, C. (2009). Social networks and health-related quality of life: a population based study among older adults. *Salud Pública de México, 51*(1).

Gaziano, T., Reddy, K. S., Paccaud, F., Horton, S., & Chaturvedi, V. (2006). Cardiovascular disease. In D. T. Jamison, J. G. Breman, A. R. Measham, et al. (Eds.), *Disease control priorities. Developing countries* (2nd ed.). Washington DC: The World Bank.

Global health council. (2008). Respiratory infections. http://www.globalhealth.org/infectious_diseases/mortality_morbidity/respiratory/. Accessed 05 May 2011.

Global Health Council. (2009). Health care workers. http://www.globalhealth.org/health_systems/health_care_workers/. Accessed 05 May 2011.

Gomez-Olive, F. X., Thorogood, M., Clark, B. D., Kahn, K., & Tollman, S. M. (2010). Assessing the health and well-being among older people in rural South Africa. *Global Health Action, 2*, 23–35.

HelpAge International. (2008). Protecting the rights of older people in Africa. http://www.helpage.org/download/4c48abf252f5f/. Accessed 05 May 2011.

Holroyd, G., Richardson, D., & Webb, J. (2005). Recasting masculinity: mapping out challenges and opportunities for public health. http://www.radcliffe-oxford.com/books/samplechapter/0347/Sexual%20Health_013-026-6a71b780rdz.pdf. Accessed 20 Jun 2011.

Hosegood, V., & Timaeus, I. M. (2006). HIV/AIDS and older people in South Africa. In B. Cohen & J. Menken (Eds.), *Ageing in Sub-Saharan Africa: Recommendations for furthering research*. Washington DC: The National Academies Press.

Ikem, R., Soyoye, D., Adebayo, J., Ojo, B., Ugwu, E., & Kolawole, B. (2010). Profile of elderly patients with diabetes admitted into Obafemi Awolowo University Teaching Hospital. *Endocrine Abstracts, 21*, 129.

Joubert, J., & Bradshaw, D. (2006). Population ageing and health challenges in South Africa. Chapter 15, 204–219. In K. Steyn, J. Fourie, & N. Temple (Eds.), *Chronic diseases of lifestyle in South Africa since 1995–2005*. Cape Town: Medical Research Council.

Kalasa, B. (2001). Population and ageing in Africa: A policy dilemma? http://www.iussp.org/Brazil2001/s80/S84_04_Kalasa.pdf. Accessed 12 Mar 2011.

Kaplan, H. S., & Bock, J. (2001). Fertility theory: Caldwell's theory of intergenerational wealth flows. http://anthro.fullerton.edu/jbock/KaplanBock_2001_FertilityTheory.pdf. Accessed 20 Jun 2011.

Khagayi, S., Debpuur, C., Wak, G., & Odimegwu, C. (2008). Socio-economic status and elderly adult mortality in rural Ghana: evidence from the Navrongo DSS. http://paa2010.princeton.edu/download.aspx?submissionId=101289. Accessed 11 Mar 2011.

Kimokoti, R. W., & Hamer, D. H. (2008). Nutrition, health, and ageing in Sub-Saharan Africa. *Nutrition Reviews, 66*(11), 611–623.

Kowal, P., Kahn, K., Nawi, N. G., et al. (2010). Ageing and adult health status in eight lower-income countries: the in-depth WHO-SAGE collaboration. *Global Health Action, 2*, 11–22.

Kowal, P., & Suzman, R. (2003). Directory of research on ageing in Africa 1995–2003. http://dc236.4shared.com/doc/np3te2Fv/preview.html. Accessed 11 Jun 2011.

Kuate-Defo, B. (2006). Interactions between socio-economic status and living arrangements in predicting gender-specific health status among the elderly in Cameroon. In B. Cohen & J. Menken (Eds.), *Ageing in Sub-Saharan Africa: Recommendations for furthering research*. Washington DC: The National Academies Press.

Kyobutungi, C., Egondi, T., & Ezeh, A. (2010). The health and well-being of older people in Nairobi's slums. *Global Health Action, 2*, 45–53.

Lamb, V. L., & Siegal, J. S. (2004). Health demography. In J. S. Siegel & D. A. Swanson (Eds.), *The methods and materials of demography*. Amsterdam: Elsevier Academic Press.

Liehr, P. R., & LoBiondo-Wood, G. (2006). *Nursing research: Methods and critical appraisal for evidence-based practice* (6th ed.). St Louis: Mosby. http://books.google.co.za/books. Accessed 15 Jun 2011.

Livesay, J. J. (2007). Cardiovascular disease in Africa. *Texas Heart Institute Journal, 34*(1), 6–7.

Lloyd-Sherlock, P. (2005). Epidemiological change and health policy for older people in developing countries: some preliminary thoughts. *Ageing Horizons, 2*, 21–24.

Lock, K., Smith, R. D., Dangour, A. D., Keogh-Brown, M., Pigatto, G., Hawkes, C., et al. (2010). Chronic diseases and development 2. http://nhfshare.heartforum.org.uk/RMAssets/NHF_Events/Lancet%20Paper%202%20-%20Health,%20agricultural,%20and%20economic%20effect%20of%20adoption.pdf. Accessed 15 Jun 2011.

Lockwood, N. R. (2003). The ageing workforce: the reality of the impact of older workers and eldercare in the workplace. *HR Magazine, 48*(12).

Lum, T. Y., & Vanderaa, J. P. (2010). Health disparities among immigrant and non-immigrant elders: the association of acculturation and education. *Immigrant Minority Health, 12*, 743–753.

Marmot, M. (2005). Social determinants of health inequalities. *The Lancet, 365*, 1099–1104.
Mbanya, J. C., & Ramiaya, K. (2006). Diabetes Mellitus. In D. T. Jamison, R. G. Feachem, M. W. Makgoba, et al. (Eds.), *Disease and mortality in Sub-Saharan Africa* (2nd ed.). Washington DC: The World Bank.
Mbewu, A. D. (2009). The burden of cardiovascular disease in Sub-Saharan Africa. Medical Research Council. http://www.saheart.org/journal/index.php?journal=SAHJ&page=article&op=viewFile&path%5B%5D=95&path%5B%5D=91. Accessed 10 Jul 2011.
Mbewu, A., & Mbanya, J. C. (2006). Cardiovascular diseases. In D. T. Jamison, R. G. Feachem, M. W. Makgoba, et al. (Eds.), *Disease and mortality in Sub-Saharan Africa* (2nd ed.). Washington DC: The World Bank.
Mills, E. J., Rammohan, A., & Awofeso, N. (2011). Ageing faster with AIDS in Africa. *The Lancet, 377*(9772), 1131–1133.
Mocumbi, A. O. H., & Ferreira, M. B. (2010). Neglected cardiovascular diseases in Africa challenges and opportunities. *Journal of the American College of Cardiology, 55*, 680–687.
Motala, A., & Ramaiya, K. (2010). Diabetes the hidden pandemic and its impact on Sub-Saharan Africa. Diabetes leadership forum. http://www.changingdiabetesbarometer.com/docs/Diabete%20in%20Sub-Saharan%20Africa.pdf. Accessed 5 Jul 2011.
Mudege, N. N., & Ezeh, A. C. (2009). Gender, ageing, poverty and health: survival strategies of older men and women in Nairobi slums. *Journal of Ageing Studies, 23*(4), 245–257.
Mwanyangala, M. A., Mayombana, C., Urassa, H., Charles, J., Mahutanga, C., Abdullah, S., & Nathan, R. (2010). Health status and quality of life among older adults in rural Tanzania. *Global Health Action, 2*, 36–44.
Nawi, N. G., Kowal, P., Kahn, K., et al. (2010). Health inequalities among older men and women in Africa and Asia: evidence from eight health and demographic surveillance system sites in the in-depth WHO-SAGE study. *Global Health Action, 2*, 96–107.
Negin, J., Cumming, R., de Ramirez, S. S., Abimbola, S., & Sachs, S. E. (2011). Risk factors for non-communicable diseases among older adults in rural Africa. *Tropical Medicine and International Health, 16*, 640–646.
Njelekela, M. A., Mpembeni, R., Muhihi, A., Mligiliche, N. L., Spiegelman, D., Hertzmark, E., Liu, E., Finkelstein, J. L., Fawzi, W. W., Willett, W. C., & Mtabaji, J. (2009). Gender-related differences in the prevalence of cardiovascular disease risk factors and their correlates in urban Tanzania. *BioMed Central Cardiovascular Disorders, 9*, 30.
Ogeng'o, J. A., Gatonga, P., & Olabu, B. O. (2011). Cardiovascular causes of death in an East African country: an autopsy study. *Cardiology Journal, 18*, 1–6.
Okunola, O. O., Akintunde, A. A., & Akinwusi, P. O. (2011). Some emerging issues in medical admission pattern in the tropics. *Journal of Dentistry Medicine and Medical Sciences, 1*, 005–008.
Pandey, M. K. (2009). Labour force participation among Indian elderly: Does health matter? http://www.crawford.anu.edu.au/acde/asarc/pdf/papers/2009/WP2009_11.pdf. Accessed 10 Jul 2011.
Rao, C., Lopez, A. D., & Hemed, Y. (2006). Causes of death. In D. T. Jamison, R. G. Feachem, M. W. Makgoba, et al. (Eds.), *Disease and mortality in Sub-Saharan Africa* (2nd ed.). Washington DC: The World Bank.
Razzaque, A., Nahar, L., Khanam, M. A., & Streatfield, P. K. (2010). Socio-demographic differentials of adult health indicators in Matlab, Bangladesh: self-rated health, health state, quality of life and disability level. *Global Health Action, 2*, 70–77.
Robinson, M., Novelli, W., Pearson, C., & Norris, L. (2007). *Global health and global ageing*. San Francisco: Jossey-Bass.
Robson, W. B. P. (2001). Ageing populations and the workforce. http://www.cdhowe.org/pdf/BNAC_Aging_Populations.pdf. Accessed 10 Jun 2011.
Sankoh, O. (2010). INDEPTH WHO-SAGE study. *Global Health Action, 2*, 2.
Smith, K. P., & Christakis, N. A. (2008). Social networks and health. *Annual Review of Sociology, 34*, 405–429.
Speizer, F. E., Horton, S., Batt, J., & Slutsky, A. S. (2006). Respiratory diseases of adults. In D. T. Jamison, J. G. Breman, A. R. Measham, et al. (Eds.), *Disease control priorities in developing countries* (2nd ed.). Washington DC: The World Bank.

Staton, J. (2008). What is the relationship of marriage to physical health? http://www.smartmarriages.com/uploaded/Staton.Health.and.Marriage.pdf. Accessed 8 Jun 2011.

Stern, R., Puoane, T., & Tsolekile, L. (2010). An Exploration into the determinants of non-communicable diseases among rural-to-urban migrants in peri-urban South Africa. *Preventing Chronic Disease, 7*(6).

Steyn, K., & Damasceno, A. (2006). Lifestyle and related risk factors for chronic diseases. In D. T. Jamison, R. G. Feachem, M. W. Makgoba, et al. (Eds.), *Disease and mortality in Sub-Saharan Africa* (2nd ed.). Washington DC: The World Bank.

Suzman, R. (2010). The INDEPTH WHO-SAGE multicenter study on ageing, health and well-being among people aged 50 years and over in eight countries in Africa and Asia. *Global Health Action, 2*, 5–7.

Takyi, B. K. (2011). Transformations in the African family: A note on migration, HIV/AIDS and poverty alleviation efforts in Sub-Saharan Africa. http://social.un.org/index/LinkClick.aspx?fileticket=p8_A7c0Xunc%3D&tabid=1555. Accessed 5 Jun 2011.

Truter, I. (2007). African traditional healers. http://www.sapj.co.za/index.php/SAPJ/article/viewFile/239/231. Accessed 5 Jun 2011.

UNDP. (2010). *Human development report 2010. The real wealth of nations: Pathways to development*. New York: Palgrave Macmillan.

UNICEF. (2010). *Progress for children: Achieving the MDGs with equity* (Vol. number 9). New York: United Nations Children's Fund.

United Nations. (2010a). *The millennium development goals report, 2010*. New York: United Nations.

United Nations. (2010b). Millennium development goals: Gender equality and women's empowerment—Progress chart. http://mdgs.un.org/unsd/mdg/Resources/Static/Products/Progress2010/MDG%20Report%202010%20-%20Gender%20Brochure%20En.pdf. Accessed 1 Jun 2011.

van Gemert, F., van der Molen, T., Jones, R., & Chavannes, N. (2011). The impact of asthma and COPD in Sub-Saharan Africa. *Primary Care Respiratory Journal, 20*(3), 240–248.

Vassilev, I., Rogers, A., Sanders, C., Kennedy, A., Blickem, C., Protheroe, J., Bower, P., Kirk, S., Chew-Graham, C., & Morris, R. (2010). Social networks, social capital and chronic illness self-management: a realist review. *Chronic Illness, 7*(1), 60–86.

Velkoff, V. A., & Kowal, P. R. (2006). Ageing in Sub-Saharan Africa: the changing demography of the region. In B. Cohen & J. Menken (Eds.), *Ageing in Sub-Saharan Africa: Recommendations for furthering research*. Washington DC: The National Academies Press.

Waweru, L. M., Kabiru, E. W., Mbithi, J. N., & Some, E. S. (2003). Health status and health seeking behaviour of the elderly persons in Dagoretti Division, Nairobi. *East African Medical Journal, 80*(2), 63–67.

Wood, R. G., Goesling, B., & Avellar, S. (2007). The effects of marriage on health: a synthesis of recent research evidence. http://njwedding.org/business/report_marriage_on_health.pdf. Accessed 15 Jun 2011.

World Bank. (2007). Demographic trends, the profile of old people, and the emerging issues. http://siteresources.worldbank.org/INTSRILANKA/Resources/LKAgingChapI.pdf. Accessed 3 Mar 2011.

World Health Organisation. (2001). Risk factors for chronic respiratory diseases. http://www.who.int/gard/publications/Risk%20factors.pdf. Accessed 6 Jun 2011.

World Health Organisation. (2002). Impact of AIDS on older people in Africa: Zimbabwe case study. http://whqlibdoc.who.int/hq/2002/WHO_NMH_NPH_ALC_02.12.pdf. Accessed 5 Mar 2011.

World Health Organisation. (2003a). Ageing and health: A health promotion approach for developing countries. http://books.google.com/books/about/Ageing_and_health.html?id=lZQpPQAACAAJ. Accessed 6 Jun 2011.

World Health Organisation. (2003b). Gender, health and ageing. Geneva. http://www.who.int/gender/documents/en/Gender_Ageing.pdf. Accessed 6 Jun 2011.

World Health Organisation. (2004). Why are chronic conditions increasing? Geneva. http://www.who.int/chronic_conditions/why/en/. Accessed 6 Jun 2011.

World Health Organisation. (2006). Determinants of health and well-being. http://www.swansea.gov.uk/hcswip/media/pdf/4/1/Determinants_of_Health_and_Wellbeing.pdf. Accessed 6 Apr 2011.

World Health Organisation. (2008a). Social determinants of health. http://www.who.int/social_determinants/en/. Accessed 6 Apr 2011.

World Health Organisation. (2008b). Urban health inequities revealed. http://www.hiddencities.org/downloads/ch4_WHO_UN-HABITAT_Hidden_Cities.pdf. Accessed 6 Apr 2011

World Health Organisation. (2009). Chronic respiratory diseases prevention and control. http://www.afro.who.int/en/clusters-a-programs/dpc/non-communicable-diseases-managementndm/programme-components/chronic-respiratory-diseases.html. Accessed 5 Jun 2011.

Zimmer, Z., & Dayton, J. (2005). Older adults in Sub-Saharan Africa living with children and grandchildren. *Population Studies, 59*(3), 295–312.

Zimmer, Z., Kaneda, T., & Spess, L. (2007). An examination of urban versus rural mortality in China using community and individual data. *Journals of Gerontology Series B: Psychological Sciences and Social Sciences, 62*(5), 349–357.

Chapter 4
Health and Health Seeking Behaviour in the Democratic Republic of the Congo: Perspectives of the Elderly

Ganzamungu Zihindula and Pranitha Maharaj

4.1 Introduction

The Democratic Republic of the Congo (DRC) is situated in central Africa and is the third largest country in Africa. A recent report indicates that the population of the DRC exceeds 70 million (World Bank 2010). The country is vast and rich in natural resources but the majority of the population lives in poverty. The DRC is one of the poorest countries in the world, with one of the lowest nominal gross domestic product per capita (UNDP 2010). Agriculture is the mainstay of the economy, contributing more than 40% to the gross domestic product. The DRC ranks 168th (out of 169) in the 2010 human development index (UNDP 2010). Poverty is a major problem in the country with 71.3% of the Congolese surviving below the income poverty line (UNDP 2010). Moreover, the latest report on the state of food security suggests that the DRC is also one of the countries that has one of the highest levels of food insecurity in the world: on average 36% of households are food insecure (WFP 2008). The Congolese Ministry of Public Health estimates that over 16% of the population suffers from severe malnutrition (MSF 2005). Hunger is a major problem for a large sector of the population. The life expectancy remains low at 48 years but there are a substantial number of older people in the country given the massive population (UNDP 2010).

For the past several decades the DRC has been characterised by severe conflict and political upheavals which have claimed the lives of millions directly or indirectly. Established as a Belgian colony in 1908, the Republic of the Congo gained its independence in 1960, but its early years were marred by political and social instability. Colonel Joseph Mobutu seized power and declared himself the President in November 1965. He subsequently changed his name—to Mobutu Sese Seko—as well as that of

G. Zihindula (✉) • P. Maharaj
School of Built Environment and Development Studies, University of KwaZulu-Natal, Durban, South Africa
e-mail: tzihindula@AfricaCentre.ac.za; Maharajp7@ukzn.ac.za

P. Maharaj (ed.), *Aging and Health in Africa*, International Perspectives on Aging 4, DOI 10.1007/978-1-4419-8357-2_4,

the country—to Zaire. Mobutu was able to retain his position for more than three decades through several sham elections, as well as through the use of brutal force. Ethnic strife and civil war, triggered by the massive influx of refugees in 1994 from fighting in Rwanda and Burundi, led, in May 1997, to the overthrow of the Mobutu regime by a rebellion backed by Rwanda and Uganda and fronted by Laurent Desire Kabila. He renamed the country the Democratic Republic of the Congo, but in August 1998 his regime was itself challenged by a second insurrection again backed by Rwanda and Uganda. Troops from Angola, Chad, Namibia, Sudan, and Zimbabwe intervened to support Kabila's regime. A cease-fire was signed in July 1999 by the DRC armed rebel groups, Angola, Namibia, Rwanda, Uganda, and Zimbabwe but sporadic fighting continued. Laurent Kabila was assassinated in January 2001 and he was replaced a month later by his son, Joseph Kabila (Amnesty International 2010; Prendergast and Atama 2009; HRW 2002). In 2003 the war officially ceased with the endorsement of peace agreements and the withdrawal of troops (HRW 2002). However, a number of reports suggest that the eastern province continues to be most severely affected by the consequences of war (Amnesty International 2010; HRW 2002). This has had serious implications for the health sector, and also has negatively impacted the health of thousands of civilians. The health of older people in the country has also suffered as a result of decades of armed conflict.

Given the situation in the country it is difficult to come across reliable data on the health situation of the population, particularly the older sector. Although very little has been written on the elderly in the DRC, a rapid assessment of the general situation of older people in the DRC conducted by HelpAge International (2009) found that almost two-thirds of older people were caring for children. Almost 14% were completely alone and 26% were with neighbours. In addition, a significant number were experiencing sight and hearing difficulties as well as physical mobility problems. More than half reported that they suffered from some form of emotional and psychological problem. In addition, 16% experienced physical abuse and 6% some form of sexual abuse. Accessibility and affordability of health care services, including medications that are necessary for the treatment of a number of chronic conditions that often afflict the elderly, such as diabetes, arthritis, high blood pressure, gastrointestinal diseases, and many others, is a major obstacle to the utilisation of services. The DRC is ranked one of the countries in which access to health care and services remains a serious challenge (UNDP 2010). A survey conducted in five provinces in the country by Medécins Sans Frontières (MSF) revealed the different and ongoing obstacles that people are facing in accessing health services. The main barriers to accessing health care services include but are not limited to: inadequate or non-existent health care provision, patients' inability to pay for health care, non-availability of quality medication, lack of supervision and training of medical personnel, non-payment of health workers' and officials' salaries, and geographical inaccessibility and non-existence of the communication structures needed for the long distances between patients' homes and the nearest health centre (MSF 2005). Another study conducted in the northern province of the DRC noted that in a situation where people were already terrorised and infrastructure destroyed by the war, it was very hard for the elderly to access health care centres (Lutala et al. 2010).

Difficulties in accessing health care are often compounded by poverty. Studies from a number of countries suggest that utilisation of basic health services consistently declines with lower socio-economic status (Szwarcwald et al. 2010).

In this context it is important to try to shed more insight into the health and health seeking behaviour of the elderly. This chapter draws on in-depth interviews conducted in both rural and urban areas of the DRC in order to obtain a detailed understanding of factors that influence the health of older men and women.

4.2 Methodology

The research was conducted in the eastern province of the DRC—the most affected by the war. While most existing literature documents the health challenges facing the elderly in urban areas, it is important to understand the differences with rural areas. Dwyer et al. (1990) point out that the rural elderly are particularly vulnerable both in terms of health status as well as access to health care services. Research suggests that rural areas are characterised by poverty, less adequate housing and transportation systems, poorer health, greater incidences of chronic health conditions, and lack of a wide range of services (Bull and Bane 1993; Krout, 1994).

The study draws on qualitative interviews with men and women aged 60 and over to understand their health seeking behaviour. Qualitative studies are useful in providing an explanation of why something is happening. Punch (2005) argues that interviews are a very good way of accessing people's perceptions, meanings, definitions of situations, and construction of reality. Furthermore, they are one of the most powerful ways of understanding others as they provide much more detailed information than what is available through other data collection methods, such as surveys. They also may provide a more relaxed atmosphere in which to collect information. People may feel more comfortable having a conversation with the researcher about the challenges they face as opposed to filling out a survey. The interviews collected detailed information about the health status of older men and women, their health seeking behaviours and some of the obstacles to accessing health services.

The qualitative data for the study comes from in-depth interviews. In total, 20 interviews were conducted with older men and women. They were selected as follows: 10 from rural areas and 10 from urban areas. For the purposes of this study, only men and women aged 60 and above were considered for participation in this study. Before the interviews were conducted, a consent form was given to the respondents. For those who could not read, the researcher or a relative read the consent form then gave it to them to indicate their willingness to participate in the study. The issue of confidentiality was highly considered and a strong emphasis was placed on it. All interviews were tape-recorded and transcribed in order to capture the responses of the respondents in their own words which allowed thorough examination of what was said (Bryman 2004). All the interviews were conducted over a period of 1 month.

To analyse the data in this study, the researcher used thematic analysis. Braun and Clark (2006) consider thematic analysis the best method for identifying, analysing, and reporting patterns (themes within data), while also organising and describing the data in rich detail and also providing the opportunity for interpretation of various research topics. This method has therefore been used to analyse data from qualitative interviews because of its usefulness in exploring contexts and meanings guided by specific themes.

The age of the respondents ranged from 60 years to 90 years. Almost half (9 out of 20) of the sample were married. The remainder was either widowed or divorced. Of the total sample, only two were divorced. Not surprisingly, women were more likely than men to be widowed. The interviews revealed that 17 out of the 20 respondents suffered from some health problems and were weak, sick, and/or physically disabled. Most of the unmarried men and women had no source of income. In the past, cultivation was their sole source of income but they were no longer in a position to cultivate the land because of ill health or the fear of violence. Out of the 10 females interviewed, four reported that they had suffered some form of sexual abuse and as a result they were afraid of venturing far from their homes. The threat of rape was a constant in their minds and as a result they were not able to go to the forest in search of food, wood or anything else. Most of the unmarried older people were dependent on their children, neighbours, and friends for their survival. The level of education of respondents was relatively low, varying from no education to secondary schooling. In total, 14 respondents had never been to school. Women were also more likely than men to have no schooling. There were only two respondents with some secondary education. The number of children per older person was relatively high, with seven of the respondents stating that they had more than eight children. Only two people reported that they did not have any living children.

4.3 Health Situation of the Elderly

Almost all the respondents reported that they were suffering from health problems at the time of the interview. They reported that they were suffering from a range of health ailments including asthma, tuberculosis, severe stomach aches, and high blood pressure. A few also stated that they were feeling unwell but were not able to explain their sickness. Only one respondent stated that they were not sick at the time of the interview or even a few days before the interview day. Some respondents observed that while they were not suffering from a particular illness their health has deteriorated over the years. They felt that they had become physically weak due to their harsh living conditions and this has rapidly turned to a sickness. Some expressed helplessness with their situation. They expressed the concern that it is unlikely that their health will improve with advancing age. Some men and women also experienced prolonged periods of illness. Both men and women who live in urban and rural areas shared similar testimonies concerning their health.

> My health continues to deteriorate. I am suffering from stomach problems, asthma and I have also become anaemic (IDI #4, rural).

> My health is very different from the way it was in the past few months because in those days I had strength, but now I am so weak (IDI #17, urban).

> My health has changed, it has deteriorated. I have been sick for the past 5 months now, and I think it will not get any better, because I am getting older (IDI #9, rural).

It is also clear that many elderly people associate old age with sickness and ill health. They expect that they will experience sickness and they feel that they cannot avoid illness. Some point out that it is not age but rather lack of food that has contributed to their illness. Hunger plagues a number of respondents. The fear of being raped serves as a major deterrent for the women to go to the forest in search of food. Men also observe that hunger has impacted their sexual life as they no longer have the desire to have sex.

> I have 34 children from 6 different wives. Even now I should still be bearing more but because of my health that is deteriorating, I feel very weak to have sex with my partners now (IDI #20, semi-rural).

> I have two wives and 16 children. We all left the town because life was hard and decided to go live in the village, but my children left the village for the town. I then became sick and both my wives cannot go out for fear of being raped. This is the reason why I can no longer even have sex with them because we both are always hungry and rarely find food (IDI #5, rural).

4.4 Health Seeking Behaviours

Respondents were asked about their health seeking behaviour, in particular the action they take when they are suffering from ill health. The in-depth interviews revealed the desperate situation of men and women who are suffering from ill health in the DRC. Their precarious financial situation places them at a great disadvantage. Just over a quarter of men and women reported that they were on medication at the time of the interview. Older people are aware that without money they will not be attended to at health facilities. Some men and women admitted that they had to stay at home until they recovered. They could not afford to go to a health facility for health care. In the urban and rural areas, cost is a major barrier to health care utilisation. Some indicated:

> Yes I did seek medication, but at times I get disappointed along the way because I know getting help from the clinic is not easy for us old people especially when you have no money to pay for your medication (IDI #16, rural).

> I did not go to the hospital, because when you do not have money you cannot be treated, and no one can pay the bill for me if I get the treatment (IDI #1, urban).

In most cases, access to treatment in the DRC is limited by financial barriers. Often citizens pay a certain amount in order to access health services. However, it is a challenge in a country with such a high unemployment rate. Some of the older people visited the pharmacy for medication when they found themselves ill. Others could not afford the costs of medication from the pharmacy and therefore sought assistance from traditional healers when their health deteriorated. In the village they often paid the traditional healer in kind rather than in cash.

> I did not go to the hospital because it is very expensive for me but I bought tablets that I have been taking here at home (IDI #6, urban).

> I did not go to the doctor because I had no money to take there with me. I went to the traditional healer who asked me to only pay a chicken (IDI #12, rural).

> I could not just go to the clinic because I did not have money. I know that no one will talk to me if I do not bring money. When the situation became very bad, I decided to go check with the traditional healer who charged me a bag of cassava which I could afford (IDI #8, rural).

Accessing health facilities in the rural areas is a huge challenge as there are few health centres in the rural areas. In addition, health centres are not close to the place of residence. The long distance to the health centre is a major barrier to utilisation of health care services. Some complained that there was no transportation to the health centres and they were therefore forced to walk long distances. Many of the rural areas in particular do not have proper roads and it is difficult to get around these areas. It often takes a long time to get to the health centre and they usually arrive only late afternoon even if they leave home early in the morning. As a result, they have to spend the night outside the health centre. In addition, the trip is long and exhausting, especially for the sick and frail.

> Yes, the hospital is situated nearly 25 kilometres away from here; it is a big problem to get there, because we have to wake up early in a morning to start the journey. We arrive there around 14:00, then after medication we sleep over there to rest a bit then wait and start the journey again the following morning. It is the same case even for those who are pregnant (IDI #15, rural).

> I walk to get to the clinic. We do not have taxis here or motorbikes to take us there, we all walk. In fact, there is no road on which to drive a car or a bike (IDI #14, rural).

> At our clinic you only get to see a doctor if you have cash in hand. Secondly, you must follow the queue and even when you are tired you can not leave because it is the only clinic that we have in the whole district and it has to cater for all of us when we are sick (IDI #1, urban).

There are also a limited number of health facilities that cater to the needs of the entire population. The few health facilities in the country, therefore, tend to be oversubscribed and there is often a long waiting period which is major problem for older people and may contribute to their deteriorating health. Both men and women who visited health facilities reported that they had spent many hours sitting in a queue in the hope that they would be seen by a health provider.

Many studies on the elderly and health care have indicated the problem of a long walking distance for the elderly when seeking medication, especially in rural areas. Some rural areas in the DRC are not accessible since they do not have any roads. In addition the journey is often time consuming and hazardous. Some elderly men and women fear visiting health facilities because of the fear of physical and sexual violence. It was not uncommon for women who were on their way to markets, churches, and hospitals to become victims of rape. In addition, in the urban and rural areas, the frail elderly who suffer from chronic illnesses do not have the physical strength to walk the distance to the health facility.

> The place where we have to go for medication is too far. The other problem is to walk in sludge and on foot. Even when it is hot the situation remains serious, and the payment is very difficult to find (IDI #7, rural).

> The health centre is not accessible, the only way I am able to get there is when I have someone to hold my hand and take me down to the hospital (IDI #19, urban).

Throughout the country, the logistical and transport infrastructure is poor which makes travel in the country extremely difficult. The situation is a result of economic mismanagement and political instability which led to under-investment in basic infrastructure. In the DRC the rural areas have been most severely affected by war. The war destroyed roads and bridges that ensured easy access to health centres.

None of the elderly reported that they were receiving any assistance from the government. They were not aware of any social grants for the elderly. The lack of government support was perceived as a major barrier to accessing health care.

> No, we have no government grant, because I have not heard of anyone receiving any form of grant since I was born. All those health insurances and pensions are new to me (IDI #3, urban).

> I have been employed by the government before but I did not even have any idea whether our government provides any form of grant such as the old age or disability grant, even the pension (IDI #4, urban).

> There is no assistance or grant that I do receive, but I would prefer knowing about it and maybe start receiving it if it exists in our country (IDI #8, rural).

Some of the men and women observed that the elderly do not enjoy much respect in society. The youth often associate the elderly with poverty, ill health, and witchcraft. They feel that they are not given sufficient recognition in society and they feel undervalued and burdensome. Many of the elderly are blamed by the youth for accelerating poverty in the community because they cannot work. They are not able to elevate their status in the community because the few jobs that they were previously involved in, farming and cultivation, can no longer be done due to insecurity in the areas. Most of the elderly, especially those who are looking after their grandchildren, reported feeling very burdened as they have no source of income yet they are expected to raise the children. The failure to obtain food and other necessities for the well-being of the children has resulted in children suffering from malnutrition and a very low rate of education. Some of those who live in urban areas indicated that they are not as affected as those living in the rural areas, because they still enjoy the support of their children and relatives in towns. But those elderly living in the rural areas, especially where all the youth have left in search of better opportunities in the towns, are worse off. Many of the elderly gave similar testimonies:

> I live alone and I am not employed. Since my two sons left for the city leaving me alone behind, I have never heard from them. They left their 3 children here and it is a problem with me to raise them. They cannot go to school because they are hungry, and there is no one to pay their school fees (IDI #11, rural).

> Life is not easy for me, I have been living in my daughter's house here in town but I always feel like I have become a burden to this couple. I left my husband in the rural area because he said he cannot afford life in the big city. Whenever I visit him, I take some food and clothing for him. In my daughter's house I am almost always the nanny or the maid.

> If there were safety in the village, I should not have left but my husband advised me to come to town to avoid those rapists, because he felt that women were more exposed than men (IDI #18, urban).

The above quotes are indicative of the differences between the urban and rural areas. Some of the elderly feel abandoned in the rural areas. In the urban areas, they often live with their children but feel that they are abused by them. They have to take on the responsibility of household chores and childrearing. Despite the issues and challenges raised by some respondents, the majority still feel a greater sense of security in the urban areas. In the rural areas the threat of rape looms heavily over women and severely restricts their movement.

4.5 Discussion

An increase in people aged 60 years and above has been observed globally. Some projections suggest that the population aged 80 years and over is estimated to increase by 233% between 2008 and 2040, compared with 33% for the total population of all ages (Kinsella and He 2009). As a result of the growing number of the older population and people living longer, there is likely to be an increasing demand for health care services. A number of studies suggest that many health problems increase with age and such a demographic trend may ultimately result in an overall increase in the number of health conditions in the population (Gijsen et al. 2001; Lutala et al. 2010). In this context it is important to better understand the health care needs of older men and women. This paper attempts to redress this imbalance by focusing on the health seeking behaviour of older men and women and the challenges they face in accessing health services.

The findings suggest that older men and women suffer from a number of health ailments. Some of the health problems tend to be more serious than others. Most of the elderly did not visit a health centre when they fell ill. The findings from the interviews suggest that this was because the elderly often do not have adequate access to health care in the DRC. There is limited coverage of health services in the DRC, especially in the rural areas. There is a huge demand on the limited number of health services.

An important issue to emerge in the interviews is how the economic situation of the elderly impacted on their health seeking behaviour. The elderly seemed particularly reluctant to seek health care because of the health costs. Many were unemployed and they did not have an independent source of income. They also did not want to burden others. In addition, they could not afford the out of pocket costs of obtaining health care. A major obstacle to accessing health care is the lack of resources. Other studies also suggest that cost is a barrier to health care utilisation. A study conducted in the DRC by Lutala et al. (2010) also highlighted financial barriers as one of the main reasons for the low utilisation of services among the poor in the DRC. In the country there is an absence of government and external funding which results in the main burden of financing health services being left to households, yet based on the poverty levels in the country, this burden is not sustainable by many households. In Pakistan, a study found that the out of pocket costs is at times as high as 80% and includes not only the consultation fee or the expenditure incurred on medicine but also the fare spent to reach the health care service (Shaikh and Hatcher 2004). Even in developed countries, there is a large body of literature which shows that people who cannot pay for health care services, either out of pocket through health insurance or public programmes or through some other means, may not receive needed health services (Bernstein et al. 2003). However, factors other than cost may also influence access to health care services.

The fear of physical and sexual violence severely restricts the movement of women, for example, going to the market, collecting water, and gathering firewood, among other activities (UNFPA 2006). In the DRC, sexual violence has been widely used as a weapon of war by the armed forces involved in the conflict. It is estimated that tens of thousands of women have become victims of sexual violence in the country. Sexual violence has been used by militia groups to intimidate and punish communities and to assert their dominance over their territory. The mass rape of women in the DRC has led to a violation of the dignity of survivors. Many choose to remain at home in order to escape intimidation and harassment. This meant that women often move to the towns to escape the violence or they remain in the village, afraid of venturing outside their homes in search of food. In this context, it is not surprising that a large proportion of the population is hungry.

The poor quality of roads, the long distances to health centres and the limited means of transport in many parts of the country, but especially the rural areas, are all major obstacles in accessing health care services in the DRC. These difficulties are largely attributed to decades of armed conflict and mismanagement first under the former leadership of President Mobutu whose government failed to invest in the health sector. After his regime came the next president Laurent Kabila in 1996 who tried to address the issues of health but was not successful. Unfortunately, while in the process, he was also assassinated in early 2002, and succeeded by his son Joseph Kabila, the current country's president, whose government has also failed due to the country's low economic situation and the unrest. This is also supported by a report which observed that "there has been a marked decline in the quality of health services in countries affected by crisis. On-going economic difficulties have undermined the public health system, resulting in an increase in the 'informal' private sector, like traditional medicine" (Commission on Social Determinants of Health 2008, p. 10).

Some elderly men and women are too frail to walk to the health facility to obtain treatment. Without the support of family and friends the frail elderly therefore cannot access health care services. Their frail state as well as their economic situation is a major disincentive to seek care, especially in the rural areas. Studies suggest that older people are most likely to suffer from chronic illnesses and also to have physical and activity limitations (Kumar 2010; Mets 1993; Lutala et al. 2010). Most reported that health facilities were situated far from their place of residence and was a long walking distance. Access to health facilities is much more problematic for the elderly frail living in rural areas than their urban counterparts. They are more likely to stay at home and hope to get better than go to a health facility. However, a few also prefer to buy their medicine over the counter while others rely on traditional medicine for meeting their health care needs.

Perception of quality of services is also likely to influence utilisation of health facilities. Both men and women complained about the interpersonal relations at health facilities. They felt that they were not given sufficient respect because they could not afford the treatment and this created some unhappiness amongst the elderly. Poor perception of health care services may lead to self-care, home remedies, and consultation with traditional healers. A number also complained that the waiting time for the consultation was unreasonably long. They had to travel long distances to get to the health facility and found that there was a long waiting time before their consultation. This meant that they could not return home the same day and this obviously placed an additional burden on them especially since some of them were frail and were suffering from ill health. They were therefore discouraged from seeking assistance at health centres because of the long waiting time. Other studies have found that long waiting times have been shown to be an important reason for the relatively high rates of programme- and method-discontinuation and may sometimes serve to discourage potential clients from seeking services, as shown by Huntington and Schuler (1993). In addition, the long waiting times at health care services may further negatively impact the health of the elderly.

As with most studies, there are several limitations. One of the limitations of the study is the reliance on self-reported behaviour. The reliance on self-reported behaviour is especially likely to be fraught with problems when the subject matter is personal and sensitive. As with any other study using retrospective data it is also likely that there were memory lapses so there may be gaps in the study. In addition, the sample was relatively small and therefore the conclusions of the study may not be generalisable to the entire population of the country. Nevertheless, the interviews shed useful insights into the health and health seeking behaviour of the elderly in the DRC.

References

Amnesty International. (2010). Amnesty international report 2010: The state of the World's human rights. London: Amnesty International. http://www.amnesty.org/en/library/info/POL10/001/2010. Accessed 20 May 2011.

Bernstein, A. B., Hing, E., Moss, A. J., Allen, K. F., Siller, A. B., & Tiggle, R. B. (2003). *Health care in America: Trends in utilization*. Hyattsville, MD: National Center for Health Statistics.

Braun, V., & Clark, V. (2006). Thematic analysis in psychology. *Qualitative Research in Psychology, 3*(2), 77–101.

Bryman, A. (2004). *Social research methods* (2nd ed.). London: University of Oxford Press.

Bull, C. N., & Bane, S. D. (1993). Growing old in rural America: new approach is needed in rural health care. *Aging, 365*, 18–25.

Commission on Social Determinants of Health. (2008). Closing the gap in a generation: Health equity through action on the social determinants of health. Geneva: WHO. http://www.who.int/social_determinants/thecommission/finalreport/en/index.html. Accessed 18 Oct 2011.

Dwyer, J. W., Lee, G. R., & Coward, R. T. (1990). The health status, health services utilisation and support networks of the elderly: a decade review. *Journal of Rural Health, 6*, 379–398.

Gijsen, R., Hoeymans, N., Schellevis, F., Ruward, D., Satariano, W., & Van de Bos, G. (2001). Causes and consequences of comorbidity: a review. *Journal of Clinical Epidemiology, 54*, 661–674.

HelpAge International. (2009). *Ageing issues in Africa. Issue No. 35*. London: HelpAge International.

HRW. (2002). *The War within the War: Sexual violence against women and girls in Eastern Congo*. New York: Human Rights Watch.

Huntington, D., & Schuler, S. (1993). The simulated client method: evaluating client-provider interactions in family planning clinics. *Studies in Family Planning, 24*(3), 187–193.

Kinsella, K., & He, W. (2009). *An aging world: International population reports*. Washington DC: US Department of Health and Human Services.

Krout, J. A. (1994). *Providing community-based services to the rural elderly*. Thousand Oaks, CA: Sage.

Kumar, N. (2010). *Health care issues of elderly most complicated*. Delhi: Varanasi.

Lutala, M. P., Kwalya, T. M., Kasagila, E. K., Watongoka, L. H., & Mupenda, B. W. (2010). Health care seeking and financial behaviours of the elderly during wartime in Goma, Democratic Republic of Congo. *African Journal of Primary Health Care and Family Medicine, 1*, 2–5.

Medécins Sans Frontières. (2005). Access to health care, mortality and violence in the Democratic Republic of Congo. Results of five epidemiological surveys; Kiwa, Inongo, Basankusu, Lubutu, Bunkeya. http://www.lakareutangranser.se/files/DRCaccesstohealthcareReport.pdf. Accessed 17 Oct 2011.

Mets, T. F. (1993). The disease patterns of elderly medical patients in Rwanda. *Central African Journal of Tropical Medicine and Hygiene, 63*, 291–300.

Prendergast, J., & Atama, N. (2009). Eastern Congo: An action plan to end the world's deadliest war. Enough. http://www.enoughproject.org/files/publications/eastern_congo.pdf. Accessed 18 Nov 2011.

Punch, K. (2005). *Introduction to social research: Quantitative and qualitative approaches*. London: Sage.

Shaikh, B. T., & Hatcher, J. (2004). Health seeking behaviour and health service utilisation in Pakistan: challenging the policy makers. *Journal of Public Health, 27*(1), 49–54.

Szwarcwald, C. L., Souza-Júnior, P., & Damacena, G. (2010). Socioeconomic inequalities in the use of outpatient services in Brazil according to health care need: evidence from the World Health Survey. *BMC Health Services Research, 10*, 217.

UNDP. (2010). *Human Development Report 2010. The real wealth of nations: Pathways to development*. New York: Palgrave Macmillan.

UNFPA. (2006). *Etat des Lieux sur la situation des violences sexuelles et les capacités des intervenants au Nord et au Sud-Kivu*. DRC: UNFPA.

WFP. (2008). Democratic Republic of Congo: Comprehensive Food Security and Vulnerability Analysis (CFSVA)/2007–2008. http://home.wfp.org/stellent/groups/public/documents/ena/wfp188266.pdf. Accessed 18 Oct 2011.

World Bank. (2010). *Democratic Republic of Congo health, nutrition and population. Country status report*. Washington DC: World Bank.

Chapter 5
Older People in Rural Ghana: Health and Health Seeking Behaviours

Nana Apt

5.1 Introduction

This chapter examines the health and non-health seeking behaviours of older people in rural Ghana. The chapter will address older people's perceptions of health and ageing and in particular the services of primary health care (PHC). The chapter will also discuss older people's reasons for not using available PHC services and also their use of alternative health care services.

A landmark of any nation's development is the health status of the people. Social determinants of health are vital to progress. These determinants are composed of the conditions in which people are born, live and mature, work, and grow old. Quite often, economics, power, and politics influence the policy choices of African nations with regard to the distribution of national resources for effective social services, especially health delivery systems. Mounting effective action to maintain the health of a nation requires both direct social policy initiatives designed to eliminate poverty and inequality, and indirect approaches. Indirect approaches are best when focused on raising public awareness about social roles and responsibilities in determining positive health outcomes, and avoiding risks from childhood to old age. Maintenance of individual health should be of paramount concern to policy makers as individual responsibility for health maintenance depends a great deal on the social contexts that condition the individual across the life cycle (Merlo 2011).

There is overwhelming evidence that social factors have a profound influence on health, and "positive health promoting influences can set in motion a health affirming cycle" (Halton et al. 2010, p. 8). This is a sure pathway leading to optimal health management in a nation. The poor in sub-Saharan Africa, including Ghana, have limited access to health services due to financial constraints, spatial differences in

N. Apt (✉)
Ashesi University College, Berekaso E/R, Ghana
e-mail: napt@ashesi.edu.gh

P. Maharaj (ed.), *Aging and Health in Africa*, International Perspectives on Aging 4, DOI 10.1007/978-1-4419-8357-2_5,

underserved areas, mostly rural, and inadequate health literacy. Indeed geographical disparities and spatial differences in Africa are important drivers to social isolation, poverty, and disease. Uneven distribution of basic infrastructure, especially road networks and communication lines, as well as remoteness from centres of trade and knowledge, work together to isolate some sections of African countries from the mainstream of development and social progress, that is, if we accept that the health of a nation is an indicator for progress. The significant rural–urban flow experienced by African countries is a direct result of this imbalance in resource allocation and social development. Various studies (World Bank 2005; UNDP 2007) confirm that excluded groups, among others, have low access to both public and private goods and services, and even when they have the services, they are of low quality. Poor educational infrastructure and health care services are typical examples. Public institutions aggravate social exclusion of segments of people living in rural areas often through lack of understanding of the linkage between exclusion and poverty and through sheer oversight. When exclusion leads to prolonged deprivation it can result in chronic poverty and chronic diseases, making old age a period of chronic illness, disability, and suffering. Besides, it can result in intergenerational transfer of poverty and disease.

In 2000, 189 countries, including African countries, signed up to the United Nations Millennium Declaration: a global commitment to halve extreme poverty and achieve equitable and sustainable development for all. Sustainable development calls for a "convergence between economic development, social equity, and environmental protection" (Drexhage and Murphy 2010). Health of a nation is therefore a key issue to be seriously considered by all nations, great and small.

5.2 Ghana's Health Promotion

Global health trends show an increasing burden of non-communicable diseases, even though in low and middle income countries like Ghana, there is a double burden of disease with high levels of communicable and non-communicable diseases. The WHO's 2004 Global Burden of Disease Survey provides evidence of the changing trends in disease patterns of the world with significant increases in non-communicable disease conditions among the general population including the older population. In Ghana, communicable, maternal, perinatal and nutritional conditions constitute 60% of all mortalities, and 6% of all injuries. Regarding non-fatal disease burden, communicable, maternal, perinatal, and nutritional conditions constitute 53%, non-communicable diseases constitute 41%, and injuries 6%, confirming Ghana's position as suffering from a double burden of communicable and non-communicable disease conditions (WHO 2007).

Ghana's key health policy objective is to improve equity and access to essential and basic health care and ensure that the health sector plays an essential role in the reduction of poverty. The current health policy framework includes the following:

- Improving geographical access to primary and emergency services by placing health points in remote rural areas
- Improving access to the financially vulnerable through the establishment of community based health planning and service zones throughout the country
- Improving socio-cultural access for priority and excluded groups such as children, women, the elderly, and people with disabilities

In Ghana, orthodox and alternative medicines operate side by side for health care, and the estimation is that about 70% of the population use alternative medicine (UNDP 2007). Alternative medicine includes traditional medicines of herbal preparations, psychosomatic treatments, and faith based medical regimes. Traditional medicine plays a vital role alongside orthodox medical practice in meeting the health needs of a vast number of Ghanaians especially those living in rural areas. It is estimated that about 80% of the Ghanaian population rely on herbal preparations for PHC (GSS 2003). There are about 50,000 traditional and other alternative medical practitioners registered in the country providing preventive and curative treatments (GSS 2003).

Orthodox health care provided by hospitals and clinics, public and private, on the other hand, lays greater emphasis on allopathic curative care and treatments as well as preventive health care (e.g. immunisation, ante-natal care, family planning, and rehabilitative services). It is estimated that about 57.7% of the Ghanaian population have access to an orthodox health facility within 30 minutes of their places of residence (GSS 2000). Urban localities have greater advantage over rural ones with respect to proximity of health facilities, better concentration of health services, adequate professional health care providers, and better services as well as better road networks. Comparatively, in many rural areas access to orthodox health facilities present challenges as residents have to travel long distances for health care often on very bad roads. This, in part, explains rural people's over-reliance on traditional health care.

Of late, health services in Ghana have made some modest gains showing improvements in some health outcome indicators such as immunisation coverage, deliveries in health facilities, and deliveries assisted by skilled health care providers (Biritwum 2006). Life expectancy at birth for Ghanaians has, however, not improved significantly over the years. Life expectancy increased from 55 years in 2003 to 57.9 years in 2006 (UNDP 2007). The factors affecting life expectancy generally in Ghana include high infant mortality, poor access to affordable and quality health care, poor access to safe drinking water, and poor sanitation, among others.

Ghana seeks to reduce by two-thirds the under-five mortality rate by 2015 in line with the Millennium Development Goals. However, child mortality remains rather high in the country. The infant mortality rate is 71 deaths per 1,000 live births, and for under five the rate is 111 deaths per 1,000 live births. This situation is the result of high levels of deprivation, poverty, malnutrition, poor access to basic education, the spread of HIV/AIDS, and the resurgence of malaria, which alone accounts for an average of 22% of all mortality cases among children under 5 years (Asante and Asenso-Okyere 2003). There are, as expected, marked differences in the probability of dying among children in urban and rural areas. Expectedly, mortality rate is higher among children in rural areas (MICS 2006; UNDP 2007). Among rural children it is 72 and 114 respectively for both infants and children under five. It is however 68 and 106 respectively for urban infants and children under five. As discussed above, and applied to the Ghanaian situation, reducing infant and child mortality will depend to a greater extent on larger investments in basic social services especially in the deprived and poor rural areas, as 65% of children under five live there (MICS 2006). Educating parents, improving nutrition and sanitation, and the provision of safe drinking water would be most beneficial to raising life expectancy.

5.3 Overview of Health Conditions in Rural Areas

It is essential at the onset to clarify several concepts. The first concept to be considered relates to the rural environment. African villages as typified in Ghana are usually built around areas with access to natural water (e.g. streams and rivers), and range in size from hamlets of a few houses as is the case in Northern Ghana, to settlements of several thousand people as is the case in the southern part of Ghana. Adequate housing has been defined as dwelling units which meet minimum building standards of safety, health, and comfort (GSS 2005). Typically in rural areas there is a growing need for better housing standards where the transformation from subsistence to cash economy is rapidly occurring.

Rural houses are usually constructed from locally available materials. In Ghana, they are single units constructed from mud, laterite/landcrete blocks, bamboo or hard wood, and roofed with thatch. Increasingly in modern times thatched roofs are rapidly giving way to aluminum roofing sheets. Villages are either composed of a cluster of single units of huts or houses surrounded by fruit trees and shrubs or farm fields. The lack of potable drinking water and adequate sanitation are also health hazards. Thus, the rural health environment is characterised by inadequate access to quality housing, potable drinking water, and facilities for waste management and disposal. These inadequacies contribute significantly to a high incidence of preventable diseases such as malaria, respiratory infections, diarrhoea, and other chronic parasitic diseases (Ackumey 2002) which affect people in their growing years to maturity. Rural Ghana is hardest hit by these inadequacies, especially in the northern part which is largely rural.

Clean water supply is vital to health promotion. Clean water and sanitation have considerable effects on reducing mortality and morbidity, and improved health depends on increasing household water supply (Gaisie and Gyau-Boakye 2007). Although access to improved sources of water has been increasing generally in Ghana, access to potable water remains a significant problem, with less than 20% of the rural population having access to clean safe drinking water (Casely-Hayford 2002). Much has been written about the differential access to safe drinking water in Ghana. For example, while 98.7% of households in Accra, the capital city of Ghana, have access to potable water, only 12.2% and 5.9% of households in rural forest areas and rural savannah areas respectively have access to treated water

(UNDP 2007). The effect of water from unprotected sources such as rivers, streams, lakes, and dug-out wells on health is much more acute among rural residents (Gaisie and Gyau-Boakye 2007) who are mostly reliant on such waters, and generally responsible for water-borne and water related diseases that are common among rural communities. An important indicator of accessibility to safe drinking water is distance. Whilst almost 98% of urban Ghanaian households have access to safe drinking water within 30 minutes of reach, 41% of rural households have access within this time and most households are not assured of a regular water supply (UNDP 2007).

Health inequities and inequalities affect individuals and sectors of communities profoundly, making rural areas more vulnerable to diseases than urban areas. Older people in rural areas who are mostly illiterate farmers are not entitled to social security or pension grants, and are among the poorest and most vulnerable in the country (Apt 1996, 2007; UNDP 2007). Not only are they poor but they lack access to requisite social services. Generally in Ghana the estimation is that the poor suffer disproportionately from disease, but their out of pocket payments for medical care and drugs lead to further impoverishment. In a SAGE sponsored health study on Ghana on Global Ageing and Adult Health, Biritwum (2006) alerted that poor households spend more on drugs than the non-poor, and that rural households spend even higher on drugs than urban households. Of interest to note in Biritwum's study with respect to discussions later in this chapter is that generally the structure of the health expenditure through out of pocket payments is higher for services such as hospital inpatient, outpatient and drugs, and much lower for traditional health care.

5.4 Overview of Ageing in Rural Ghana

The demographic profile of Ghana indicates that people are living longer and will continue to live longer in the years ahead. However, Ghana's population is not only youthful as a whole but it is so in every subsection. The median aged worker is only 30 years and one member in nine of the labour force is over 50 years of age. Nevertheless, over the years, Ghana's older population, 60 years and over, is showing signs of rapid growth (see Table 5.1). Between 1960 and 1970, Ghana's population aged 60 years old and above increased by 8.2%. From 1970 to 1984 the increase was 9.4%. Thus, within a period of 24 years the older population group increased by 18.4%, indicating that Ghana's population is growing older.

Table 5.1 Population of Ghana aged 60 years and above

Year	Absolute	As a % of total population	Period	Increase
1960	331,516	4.6	1960–1970	8.2
1970	457,873	5.3	1970–1984	9.4
1984	719,135	5.8	1960–1984	18.4

Computed using data from the 1960, 1970, and 1984 Population Censuses of Ghana

Table 5.2 Projected geographical distribution of 60+ population

	60–64	65–69	70–74	75–79	80+	Total
1984 Urban	58,327	38,658	32,807	18,072	32,062	179,926
1984 Rural	167,449	106,655	96,056	53,741	115,309	539,210
1991 Urban	82,367	52,412	44,484	24,504	43,474	247,241
1991 Rural	200,874	126,695	15,233	64,468	138,326	545,596
2000 Urban	120,789	80,048	67,940	37,425	66,397	372,599
2000 Rural	236,957	115,927	135,933	76,048	163,173	728,038
2025 Urban	262,184	173,752	96,556	81,235	144,121	757,848
2025 Rural	247,240	157,477	141,832	79,348	170,254	796,151

Computed using data from the 1984 Population Census of Ghana

Table 5.2 shows the geographical distribution of the 60 years and over age group. For all the computed years, there are more older people living in rural areas until about 2025. The projections suggest that in 2025 there are going to be slightly more older people in urban areas than in rural areas.

In general, the majority of Ghanaians still live in rural areas although the proportion of the population living in urban areas is increasing. Census data indicate that the urban proportion of the total population rose from 23% in 1960 to 32% in 1990, the result of rural–urban migration combined with high fertility. The link between rural–urban migration in Ghana and the ageing of rural areas has been well documented. In the movement to Ghana's cities and towns from rural areas, the youth dominate (Caldwell 1967; Addo 1972; Nabila 1986; Engman 1986). The drift from the rural areas to urban areas is therefore contributing to the isolation of older people in rural areas as well as depriving them of their usual sources of social and economic support (Apt 1985). Migration from rural areas to urban areas by the young affects the care and well-being of older people at three levels. Firstly, the departure of resource persons in the family and household generally is devastating, as is, secondly, the departure of caregivers of the old, mostly women and, thirdly, the inability of the migrant family (mostly children) to provide older people left behind with adequate income for their needs, due to increasing unemployment and low salary levels even for the employed. In addition, income security of older people in Ghana has been found to diminish with age (Okraku 1985; Apt 1985, 2007) as a result of low pensions for those retiring from formal employment and the absence of pension schemes and social security benefits for those retiring from informal work (mostly peasant farmers, fishermen, artisans, and market traders). It is estimated that about 60–80% of the working population are engaged in subsistence farming and other non-formal work. Due to women's even weaker participation in organised employment they are less in a position to receive retirement benefits. Older women are often widowed or divorced and are more likely to live alone. The greater majority of rural older people thus belong to the informal sector. The traditional expectation that the younger generation take care of ageing members of the family when they have passed their working years no longer holds, mainly due to modern economic constraints and the high cost of living. Consequently, an increasing number of older people are resource neglected. Children constitute the highest percentage of income sources for older people in both areas but especially for the rural. Daughters, sons, grandchildren, and siblings in this order are the main contributors to elderly welfare (Apt 1989), and while older women receive more welfare from their children, they are however less well off in old age compared to men. Older men, widowed, remarry, and have more chances of being cared for by a partner, whereas older women, widowed or divorced, are likely to remain unmarried.

Loneliness and dissatisfaction appear to be the emerging climate for older people left behind in rural areas. From separate studies in two regions of Ghana, Barnie (1995) and Twum (1993) have attributed loneliness of older people to infrequent visits by their children, but older people who are childless are especially lonely.

Health begins with well-being. Are older people satisfied with their life? In a study of the Central Region of Ghana (Apt 1989), a significant association was found to exist between self-assessed life satisfaction of older Ghanaians and the following variables:

- Type of accommodation
- Having children
- Living with family members
- Involvement in community affairs

A very high proportion of older people living in their own house reported satisfaction with life whereas the opposite was expressed by the majority in rented accommodation which was more common in urban areas. Although the number of respondents who were childless was quite small (5%), the majority (66%) expressed dissatisfied feelings. The percentage of this group who were urban residents was higher (77%) compared to their rural counterparts (45%). Although there was no evidence showing that older people involved with their community in general have higher life satisfaction, it was highly significant that respondents who were members of community associations/organisations appear to be more satisfied with life. The analysis similarly showed a highly significant relationship between life satisfaction and living with family members. Emerging from this study, and being of importance for programme planning for older people, is the expressed need of older people both in rural and urban areas to be with family, and to be of need to them (children, grandchildren, spouse, and siblings). In addition to expressed need for family links and social interactions with the family network, older people have at all times listed financial support, healthcare, and housing as necessary preconditions for well-being (Brown 1984; Okraku 1985; Apt 1994). In the Central Region study referred above, those expressing life satisfaction gave two distinct reasons (a) having good care (60.4%) and (b) having money/property (71%). Of those expressing extreme dissatisfaction with life, economic reasons topped the list of grievances (73.5%).

Health of older people, particularly in rural areas, lacks comprehensive data in Ghana. However, a national survey of the elderly in Ghana (Apt 1993) indicates that older people in rural areas perform better than their urban counterparts in activities of daily living. However, rural older people compared to urban ones perceive their health as being either poor or very poor. Of their assessed health problems the following in order of most frequent occurrence were recorded:

- High blood pressure
- Heart attack
- Stomach ulcer
- Arthritis/rheumatism
- Foot problems
- Falls
- Lung disease
- Stroke/paralysis

We can therefore conclude that in general, older people in Ghana suffer from multiple health problems. This factor has received confirmation from the findings of a 1995 study of Ghana's Ministry of Health at Nsawam in the Eastern region of Ghana (Banga 2000). Rheumatism/arthritis and joint pains, eye problems, and hypertension were especially listed as current ailments and most of the older persons in the sample had these medical problems for long periods prior to the interview. Besides and of interest to the main discussion in this chapter is the fact that even though a majority of the older people interviewed (78%) did go to hospitals/ clinics when ill, distance, lack of funds, and long periods of waiting for medical attention deterred their frequent attendance.

5.5 Objectives and Methodology of the WHO INTRA Study

WHO developed a project entitled "Integrated Health Care Systems Response to Rapid Populations Ageing in Developing Countries—INTRA", which focused specifically on the response of the PHC sector to population ageing. The WHO INTRA study in Ghana, which forms the basis for discussion in this chapter, was designed to investigate the health and non-health seeking behaviours of the non-users of the PHC services in Ghana. Five urban and rural localities were selected to reflect the different infrastructure and social contexts in rural and urban Ghana. In this respect Aseseeso, Ashiyie, and Twepease rural localities in the Greater Accra and Eastern regions of Ghana were purposefully selected to reflect differing rural population sizes and access to PHC services whilst the two urban communities in the study, Mamoobi and Awudome, suburbs of the city of Accra, the capital, were selected using the simple random sampling procedure. Utilising only the focus group discussion methodology and WHO protocols assigned, group discussions were held in local languages in separate groups of men and women in each locality to ensure full universal participation of group members. The discussions in each community lasted between 90 and 120 minutes with recordings made of the discussions following strict guiding WHO protocols for the study. Participating members of the focus group discussion ranged from 50 to 75 years.

5.5.1 Brief Profile of Selected Rural Localities

5.5.1.1 Twepease

Twepease, located in Ghana's Eastern region near the district capital, Nkawkaw, has the lowest population size of 183. Contrary to all the selected sites (urban and rural) Twepease has a higher male population (98 men and 86 women). Twepease village has a primary school and a junior secondary school. The nearest senior secondary school is about 30 kilometres from the village. Twepease has no public hospital but a small clinic run by the Salvation Army. It also has a private facility for traditional healing.

5.5.1.2 Aseseeso

Aseseeso is a rural community on the Akwapim Ridge in the Akuapem North District also in the Eastern Region. Aseseeso is surrounded by farming communities such as Abonse, Totease, Abokyi and Akuni. The total population size for the Aseseeso locality is 970, made up of 429 males and 541 females. The number of households found within the Aseseeso community is 256. These households live in 130 houses with an average household size of 3.8 persons. Like Twepease and most rural areas in Ghana, Aseseeso lacks many social amenities such as hospitals and schools. Aseseeso has however a primary and junior secondary school. The nearest senior secondary school is 40 kilometres away. Aseseeso, like Twepease, lacks a public hospital facility. It has a private clinic run by a non-governmental organisation which caters for health delivery services. In addition, there is a traditional healing centre.

5.5.1.3 Ashivie

Ashiyie, the third rural locality in the Greater Accra region selected for the INTRA project, is located on the Dodowa road, near Adenta residential area, east of the city of Accra. Ashiyie has a population size of 534, made up of 281 females and 253 males. The total number of households in Ashiyie is 114 living in 82 houses. The average household size is 4.7 persons. Unlike all the project localities, Ashiyie has no primary school, neither has it a junior secondary school nor a senior secondary school. This may be due to the proximity of the locality to Adenta which has good educational facilities. Like most rural localities, Ashiyie has no hospital and like the two other selected villages, it has a private not for profit clinic and a traditional healing facility.

5.6 Findings

5.6.1 Basic Characteristics of the Rural Participants

Almost 99% of the participants were either married or widowed. The majority of the male participants were married compared to the women who were either widowed or divorced/separated. In Aseseeso for example all the male participants were married while in Ashiyie a significant number of the men (52%) were widowed. The education background was low for both males and females but more so for the women who were mostly illiterate and primary school leavers. Two women in Aseseeso completed secondary school and four in Twepease and Ashiyie had some secondary education. With regard to employment, the majority, 62%, were small-holder farmers and the rest were petty traders. Two male participants in Aseseeso, who had completed tertiary education, were retirees from government work.

Self-assessed current health status as compared to the urban groups was poorer. The majority of the male and female participants did not report excellent health status but viewed their current health as mainly fair. Women on the whole had a poor perception of their health status as only a few mentioned that their health condition was good.

Self-reported health problems or diseases ever suffered or still suffering from were as follows:

- Arthritis
- Asthma
- Diabetes
- Hypertension (reported more by women)
- Stroke (reported more by men)
- Chronic lung diseases (reported more by men)
- Eye diseases
- Skin diseases
- Malaria
- Injuries (reported more by women)

Overall, the majority of the participants (63%) lived in single rooms in compound houses with children and other family members. Others, occupied single rooms in family houses (30%) and 7% lived in rented rooms. There were significant numbers of children and other relatives in each household. The household composition shows to some degree the availability of family support. There were significant numbers of children and other relatives in each household. The household structure in old age may reflect, on the one hand, socio-cultural norms or preferences relating to family ties and co-residence—such as a preference for extended family living among the older adults. On the other hand, the residential distribution between the localities may reflect resource limitations. Such constraints may necessitate joint extended family residence, as living in separate households may not be affordable.

5.6.2 Older People's Perception of Health and Well-Being

Seeking health care in old age was perceived by the focus group discussants as "necessary", "important", and "vital" to maintaining good health, but financial constraints limited frequent visits to hospitals and clinics. On the whole the perspective was that it would be good to have regular medical check-ups. Women believed a monthly check would be ideal and the men thought four monthly periods would do. Both female and male participants across the spectrum of group discussions were vociferous in their conceptualisation of well-being in old age. The following comments clearly emphasise their need for a resource base.

Well-being means:

- Living well
- Having a good balanced diet
- Leading a comfortable life
- Being financially able to take good care of oneself
- A lack of financial worries

Some comments made by the participants with respect to the above are as follows:

> Life wouldn't be that bad if I lived in my own house. I must pay rent for my lodgings. How can I eat well and feel good?

> My children are doing their best to provide me with money but life is hard for them too. I must be content with what they can provide.

> If I had only known life would be so expensive!

Others perceived close family ties as an important indicator of well-being:

- Living with family
- Having children who care about you
- Ability to take care of my grandchildren
- Being consulted by my children

Some comments made by the participants with respect to the importance of close family ties are as follows:

> I am living for my grandchildren who are living with me now.

> What has this life come to? My son married without consulting me. He brought his wife to see me when the marriage was already concluded. Do you call this act respectful?

Physical and psychological fitness also featured in their responses to well-being:

- Having no problems
- Doing what young people do including sexual activities
- Physical fitness
- Being active
- Being able to work on my farm
- Lack of worry
- Being respected

Some of the participants in the group discussions, especially men, were found to excessively smoke and take alcoholic beverages of the strongest type brewed locally (known in Ghana as *"akpeteshi" or "kill me quick")*. A male participant from Aseseeso remarked, "Alcoholism and chasing of women make people weak. The opposite makes one stronger even at old age".

To sum up, health and well-being as discussed have to do with the "good life". The good life in the perception of the older people is having enough to eat, having family around and having money to seek medical care when required, but above all "having respect".

5.6.3 Use and Non-use of PHC Services

Older people in rural areas are aware of hospital services but cannot afford the cost. Some would attend the hospital but not fill the prescription given them because they did not have sufficient funds or they would purchase part of the prescription if there were more than one drug prescribed by the doctor and tell the pharmacist they would come back for the rest but never return. Many others would make the occasional visit to the hospital/clinic and discontinue visits due to their inability to pay. Above all else they showed distrust and anxiety about going to hospitals. Generally, a lot of concern was raised about the manner older persons were handled at hospitals. More and more participants of both sexes stated during the discussions that they would go to the hospital as a last resort when all else had failed. All else, in this respect, would mean self- medicating with purchased drugs or herbs, seeking treatment at a herbal centre and/or seeking spiritual healing. Attendants at hospitals usually located miles away in urban areas received the most complaints. They complained generally about long queues and waiting hours to assess services. They complained about the lack of consideration and respect from hospital medical personnel and ill treatment especially by nurses. The prohibitive cost of prescription drugs was above all else a great deterrent factor for utilising primary health services.

> I could not go back to the hospital even though the doctor asked me to. I spent all my money for one visit and I couldn't even pay for all the drugs prescribed.

> The nurses have their own favorites when you are in the queue. They will make you stand in line till your legs hurt and if you complain they will insult you.

5.6.4 Health Seeking Alternatives

Alternative care for many was simply self-medicating. That is, buying directly from a community drug store what they considered or had heard from friends and relations as good medicine or buying from drug peddlers on their recommendation of what is good for their particular ailment. In Ghana, very often drug peddlers go to rural areas to sell drugs to rural people with convincing tales of the power of the drugs they are selling. Much of the drugs are harmless vitamins or aspirins, others are prescription drugs that could be quite dangerous to take if not prescribed by a doctor. These peddlers take advantage of the illiteracy and ignorance of the rural people (Banga 2000). Quite a number of old people also self-medicate with herbs they know or have used all their growing years or get prescription herbs from traditional herbal centres. As a matter of fact all participants irrespective of gender use herbal medicine, some more frequently than others. They found herbs and self-medication to be cost effective. There were also some older people who combined self-medication either with drugs or herbs with spiritual healing.

> When I am sick I inform my pastor and he prays for me. Sometimes I attend the clinic but my pastor always prays for me to heal quickly.

> I go to the herbalist when I don't feel fine and get some herbs. Herbs don't cost much. It has helped me all these years.

> When I have not enough money to go to the clinic I buy drugs from the drug store.

> Sometimes I buy drugs from the drug seller. He shows me what good drugs to buy.

5.7 Emerging Implications for Policy

The main goal of health promotion is to promote and maintain good health. Does Ghana meet this goal from the perspectives of older people in rural areas? The focus group participants did not feel healthy and bemoaned the fact that they lacked resources to benefit from primary health services. From the villages many travelled to hospitals at far distances which added to the dilemma of not having enough money to pay for hospital drugs. Improvements in the health of the older people in Ghana will require increased access to health care and health education. To some degree, the diseases that afflict Ghana's older people are amenable to prevention or modification through changes in behaviour (such as cutting down on smoking and alcohol usage, or exercising) and enhanced access to health care services (for example, screening for hypertension and diabetes).

In general, the study suggests that older persons from rural communities tend to engage in unhealthy behaviour more than their urban counterparts, e.g. smoking and alcohol use. Therefore, understanding why persons of low socio-economic status tend to engage in unhealthy behaviour more than those of high socio-economic status and to use health services less than them is a prerequisite to improving the health of the aged in deprived communities in the country.

The government should be committed to increasing the quality and years of healthy life for all Ghanaians irrespective of where one lives. Lack of money was a recurrent theme in this study as to why PHC services are not utilised despite knowledge of it. Poor health and poverty are strongly correlated. All the male participants from Tweapease, for example, who complained of poor health declared "We do not have money to go to the clinic" as the first reason for non-use of PHC services. Indeed, all the participants from the five study areas (rural and urban) favoured some kind of concessional provisions for the aged to enable them to access health services. In this respect, the National Health Insurance Scheme, currently being implemented in Ghana with a view to offering affordable medical care, especially to the poor and vulnerable, should address the issue of free or subsidised medical care for persons aged 60 years and above. Currently it is free for 70 years and over, while age 60 is the mandatory retirement age in the country.

To some extent, poor health may cause poverty by restricting the hours an individual can be active and the kinds of work he or she can perform, as well as by necessitating costly medical care and special equipment and services. Two male participants from Aseseeso rural community lamented:

> I am finding it very difficult to work on my farm. I am no longer able to make the money I used to make.

> As for me I think I am not healthy because there is no money to buy proper food and obtain medical care.

A woman from Ashiyie locality stated, "I used to farm 10 years ago but now my sight is disturbing me and I can't even walk a little distance". Asked why she did not go to have her eyes checked at the hospital, she stated simply—"I can't afford it". Furthermore, when a male participant from Tweapease said, "My legs are not as strong as before", all the other members of the focus group nodded strongly in agreement.

The trends noted above, with all things being equal, indicate the desperation of rural people with respect to health matters. It is obvious that there is a need for a paradigm shift in the way health services are provided and the shift in paradigm needs to focus strongly on achieving the policy frameworks set by the Health Ministry. In particular the following need an urgent focus (Apt 2007):

- Improving equity in access to basic health services
- Improving efficiency and responsiveness to the needs of all people in Ghana
- Developing effective inter-sectoral collaboration

Consequently, policies targeted at enhancing health and improving access to health services must acknowledge that over the life cycle, poverty may both influence health and be influenced by it. The number of health facilities available need to be doubled and the bulk of the facilities should be situated in rural districts where the majority of Ghanaians, old and young, reside.

5.8 Conclusion and Recommendations

Traditionally, older people constitute an integral component of Ghanaian society. Their importance and power base might be eroding under modern circumstances. Nevertheless, in the context of increasing numbers and their inherent vulnerability especially those living in rural areas, the concepts of PHC with its wide scope of activity and influence appear particularly relevant in defining and meeting older people's health needs. In the area of health promotion the following recommendations are made with respect to PHC. Ghana should of necessity:

- Assure safe drinking water, adequate food supplies, and sound nutrition
- Reduce the possibilities of accidents, causing disabilities in high risk areas particularly on roads
- Promote early diagnosis and appropriate treatment as well as preventive measures to reduce disabilities and avoid premature ageing
- Encourage exercise and other elements of a healthy lifestyle
- Ensure universal and affordable access to adequate health care

For monitoring older people's health and early identification of those likely to be at risk no matter their geographical residence, community clinics linked to the Maternal and Child Health (MCH) clinics which are well rooted in the country's health planning, can provide easily accessible service and represent a form of preventive medicine for old people. In addition, services rendered by the Department of Social Welfare under the Community Based Rehabilitation (CBR) should be expanded to include older persons with disabilities and, for the bedridden that require extensive medical attention, trained community volunteers can be an invaluable source of community welfare.

HelpAge Ghana, the first and most active age care organisation in Ghana, and other not-for-profit organisations, including religious bodies could take up this challenge of volunteer training as currently in Ghana, there are many men and women as well as youth groups and associations that could be challenged through training to take up age care volunteer leadership. Additionally, the development of health education and services in schools and colleges through which young people acquire health knowledge would enable the youth to become health activists in their communities and to promote age care in rural communities especially for their required national service.

An increase in chronic illness and long-term disability is related to increased lifespan. Therefore, it is necessary for Ghana to develop appropriate guidelines in response to older people's health needs. A changed emphasis in the provision of personnel training, specialisation and services to include the needs of the increasing population of older people is a major area for government action. The needs of mothers and children will remain but those of a growing number of older people must be recognised, planned for and incorporated in the regular mother and child health schemes that currently operate.

Regular utilisation of hospitals and clinics are problematic due to the poverty of older people particularly in the rural areas. Fortunately, Ghana, since 2007, has introduced a national health insurance system which waves off contributory payment other than registration fee for children and people 70 years and above. The indication is that the poor including older people in rural areas cannot afford the basic registration fee, so they do not register. The health insurance scheme is under review presently and hopefully all bottle necks which have come up for discussion will be smoothed out to enable greater use of the scheme by older people.

Family and intergenerational networks appear to show significant landmarks for maintaining the health and well-being of older people in Ghana especially in rural areas where "abandonment" is most likely to occur. As earlier discussed, social factors have a positive influence on health. The importance of social factors in the determination of well-being is prominent in the discussion of the WHO study as well as earlier studies on ageing cited in this chapter. Many older people form an integral part of the family structure even now in Ghana and, thus, intergenerational relations and exchange are very important indicators to elderly well-being. Policy makers and age care planners will need to be sensitive to these aspects of family life in particular and having specific policies in place will encourage family integration. For instance, day centres for older people would enable families to share the responsibility for the care of older family members, a sharing that encourages continued family support while relieving them of some of the burden of caring. HelpAge Ghana and its affiliate organisations have opened 3 day centres in Accra. Such centres should be opened in rural areas as projects of District Assemblies and the Department of Social Welfare.

Special attention needs to be paid to the significance of loneliness, isolation, and family dispersion in so far as these factors may readily lead to secondary consequences of apathy in old age. Family members must be seen as prime providers of support for older people and therefore in need of education and training in the best methods of caring for their elderly members and in some respect, financial support to do a good job.

References

Ackumey, M. (2002). Local perceptions of Buruli ulcer in the Ga district, Greater Accra Region. *Social Policy: Journal of the Centre for Social Policy Studies, 2*(2), 44–57.

Addo, N. O. (1972). Urban population and employment in Ghana. In S. Ominder & C. N. Ojuoku (Eds.), *Population growth and economic development in Africa*. London: Heineman.

Apt, N. A. (1985). The role of grandparents in the care of children. In P.A. Twumasi (Ed.), *Problems and aspirations of Ghanaian children: Implications for policy planning and action*. Project Report for Ghana National Commission on Children. Sociology Department, University of Ghana, Legon.

Apt, N. A. (1989). *Impacts and consequences of social change on ageing in Ghana: A study of the central region*. Unpublished doctoral dissertation. University of Ghana, Legon, Ghana.

Apt, N. A. (1993). *Survey of the elderly in Ghana. Report for United Nations Office*. Vienna: Centre for Social Development and Humanitarian Affairs.

Apt, N. A. (1994). *The situation of elderly women in Ghana. Report for the United Nations*. New York: United Nations.

Apt, N. A. (1996). *Coping with old age in a changing Africa*. Aldershot, UK: Avebury.

Apt, N. A. (2007). *Learning how to play to win*. London: Motjuste.

Asante, A. F., & Asenso-Okyere, K. (2003). *Economic burden of malaria in Ghana*. Technical report submitted to the Africa Regional office of WHO. Harare.

Banga, E. (2000). Emerging images of the aged in Ghana: Health and nursing implications. Paper presented at a seminar on Nursing Education in the 21st Century. Department of Nursing, University of Ghana, Legon.

Barnie, L. (1995). Ageing in Brong Ahafo region. Unpublished dissertation, University of Ghana, Legon, Ghana.

Biritwum, R. (2006). *A study of global ageing and adult health*. Ghana National Report. SAGE sponsored study (Unpublished).

Brown, C. K. (1984). *Improving the social protection of the aging population in Ghana*. Legon: University of Ghana, ISSER.

Caldwell, J. C. (1967). Migration and urbanization. In W. Birmingham, I. Neustadt, & E. N. Omaboe (Eds.), *A study of contemporary Ghana* (Vol. 2). London: Allen and Unwin.

Casely-Hayford, L. (2002). The social impact of poverty of educational achievement: a national perspective. *Social Policy: Journal of the Centre for Social Policy Studies, 2*(2), 1–17.

Drexhage, J., & Murphy, D. (2010). *Sustainable development: From Brundtland to Rio 2012. Background paper for the high level panel on global sustainability*. New York: United Nations.

Engman, E. V. T. (1986). *Population of Ghana 1850–1960*. Legon: Ghana University Press.

Gaisie, S., & Gyau-Boakye, P. G. (2007). Population growth, water/sanitation and health. In C. J. Mba & S. O. Kwankye (Eds.), *Population, health and development in Ghana* (pp. 91–93). Accra, Ghana: Sub-Saharan Publishers.

Ghana Statistical Service (GSS). (2000). *Ghana living standard survey report*. Accra, Ghana: Government of Ghana.

Ghana Statistical Service (GSS). (2003). *Core welfare indicators*. Accra, Ghana: Government of Ghana.

Ghana Statistical Services (GSS). (2005). *2000 Population and housing of Ghana*. Accra, Ghana: Government of Ghana.

Halton, N., Larson, K., & Russ, S. (2010). Why social determinants? *Healthcare Quarterly, 14*(Supplement), 8–12.

Merlo, J. (2011). Contextual influences on the individual life course: building a research framework for social epidemiology. *Psychosocial Intervention, 20*(1), 109–118.

MICS. (2006). *Monitoring the situation of children, women and children*. Accra, Ghana: Multiple Indicator Cluster Survey.

Nabila, J. S. (1986). Rural urban migration and its implication for rural development in Ghana. In C. K. Brown (Ed.), *Rural development in Ghana*. Legon: Ghana University Press.

Okraku, I. O. (1985). A study of Ghanaian public service pensioners. Paper presented at the XIIIth International Congress of Gerontology, New York, 12–17 July.

Twum, S. K. (1993). *Health of the elderly in the Eastern region of Ghana*. Accra, Ghana: Ministry of Health.

UNDP. (2007). *Ghana human development report*. Accra, Ghana: UNDP.

World Bank. (2005). *Pro-poor growth in the 1990s lessons and insights from 14 countries*. Washington: World Bank.

World Health Organization (WHO). (2007). *National health accounts: Country information*. Geneva: WHO.

Chapter 6
Growing Old in the Era of AIDS: Perspectives and Experiences of Older Men and Women in South Africa

Chantal Munthree and Pranitha Maharaj

6.1 Introduction

Like many other developing regions, South Africa is currently experiencing a growth in the number of older people. South Africa has one of the most rapidly ageing populations in Africa with more than one in eight persons aged 50 and over and nearly 7% aged 60 and over (Statistics South Africa 2011). The decline in fertility is resulting in a steady shift towards an ageing population. South Africa has been one of the first countries in Sub-Saharan Africa to experience an overall fertility decline (Caldwell and Caldwell 1993). The overall fertility rate has fallen from 6.0 in the mid-1950s to about 4.3 in the 1980s and is currently 2.3 (Caldwell and Caldwell 1993; Statistics South Africa 2011). However, a certain degree of variation exists across geographical areas and racial groups. Despite this, the fertility rate in South Africa is much lower than those recorded elsewhere in Africa. The current fertility rate of 2.3 is expected to decline further. This trend is likely to have major implications for the age sex structure of the population, leading to a decline in the younger population and a growth of the older population. In the past 5 decades the share of the population aged 60 and above has increased steadily to 3.28 million and is projected to rise to 10.5% by 2025 (Joubert and Bradshaw 2006). The AIDS pandemic, which increased from less than 1% in 1990 to almost 28% in 2007 according to the national antenatal survey, is also likely to influence the age structure of the population of South Africa (Department of Health 2008a).

The opportunity cost of AIDS on development prospects is catastrophic. South Africa is a middle income country with one of the highest Gini coefficients in the world. In 1993, the Gini coefficient for South Africa was 0.66; it rose to 0.70 in 2008, suggesting that income inequality is worsening (Leibbrandt et al. 2010). This

C. Munthree (✉) • P. Maharaj
School of Built Environment and Development Studies,
University of KwaZulu-Natal, Durban, South Africa
e-mail: munthreec@ukzn.ac.za; Maharajp7@ukzn.ac.za

P. Maharaj (ed.), *Aging and Health in Africa*, International Perspectives on Aging 4,
DOI 10.1007/978-1-4419-8357-2_6,

places South Africa in the position of one of the most unequal societies in the world. The full and most devastating impact of AIDS on South Africa will not be entirely reflected in its macro-impact. The full scourge of the pandemic is impacting most negatively on human development indicators, more specifically life expectancy. Between 1995 and 2002, life expectancy declined from 61.4 years to 51.4 years (UNDP 2003), but recently there has been a slight improvement, probably as a result of wider access to antiretroviral treatment. In South Africa, AIDS primarily affects adults in the prime working age groups shifting the usual patterns of mortality and morbidity (Shisana et al. 2005). With the introduction of antiretroviral therapies tremendous progress has been made in the treatment of AIDS. However, these drugs remain largely out of reach for a significant proportion of people living with HIV/AIDS (Maharaj and Roberts 2006). By the middle of 2006 only 46% of people needing treatment enrolled in the antiretroviral treatment programme, and 36% were initiated on the antiretroviral treatment programme. In 2007, of the 889,000 people that needed treatment, 55% enrolled and 42% were initiated on the antiretroviral treatment programme (Department of Health 2008b).

Over the past decade mounting evidence suggests that older people in Africa are increasingly becoming infected by HIV/AIDS. For instance, infection data from Uganda collected between 1992 and 2002 shows that the number of people over the age of 50 registering for HIV testing and counselling services rose sharply from 3% to 30% and of these, a fifth have tested HIV positive (HelpAge International 2004). Although sexual activity may decline with age, it is suggested that it is still commonly regarded as an important risk factor for the spread of HIV infection among older adults (Best 2004; Knodel et al. 2002). Early symptoms of HIV infection can and are often mistaken for signs of diseases associated with ageing, preventing those infected from seeking appropriate HIV treatment (UNAIDS 2002). Older people may also experience chronic and acute illness commonly associated with ageing such as hypertension, arthritis, and diabetes, and the added burden of HIV/AIDS may make them increasingly vulnerable to an early death (Knodel et al. 2002). Infection figures of older people do not reflect at what age the infection occurred, but it is highly probable that many of those over 50 were infected at a younger age and so many older AIDS sufferers may have had the virus for many years before being tested, and as a result, may find themselves at an advanced stage of the infection in their old age (UNAIDS 2002). Due to the delay in diagnosis and treatment, and the fact that age seems to accelerate the progress of HIV to AIDS, older people who are infected do not survive as long as younger people (US Centers for Disease Control and Prevention 1998). Given that little is known about the risk of HIV infection among older people, this study will fill a significant gap in the literature.

Older people in Africa are also directly impacted by the AIDS pandemic. In many parts of Africa, older people have traditionally been financially taken care of by their adult children; however, as a result of the HIV/AIDS pandemic there is a complete reversal of intergenerational roles (HelpAge International 2004). The more productive members of the family who are in their prime adult years have the highest levels of infection and, as a result, are more likely to experience increased morbidity and mortality. In many parts of Africa, older people often assume primary responsibility for the health and well-being of sick family members in the absence of effective government social welfare systems (Moore and Henry 2005). Surveys conducted in seven countries in Sub-Saharan Africa show that approximately 40% of people living with HIV were being cared for by older people (HelpAge International 2006). A number of studies in Africa suggest that older family members represent a critical source of support for chronically and terminally ill HIV infected children (Ntozi and Nakayama 2001; Ssengonzi 2007). Often adults who become sick with AIDS related illnesses turn to their parents when they are no longer able to manage by themselves. A study conducted in six districts in Uganda found that parents were often the primary caregivers of AIDS patients (Ntozi and Nakayama 2001). Increasingly, studies also show that older people take responsibility for the care of their orphaned grandchildren (HelpAge International 2006; Mall 2005). In Namibia, South Africa, Botswana, Malawi, Tanzania, and Zimbabwe, almost 60% of orphaned children reside in grandparent-headed households (Mall 2005). In many instances, grandparents provide care for several grandchildren. Older women are more likely than older men to assume responsibility for caregiving activities (Mall 2005; Chazan 2008). This informal caring is provided in the context of high levels of poverty, lack of support, stigma, and accusations of witchcraft as well as many of the other challenges frequently confronting people in their old age (May 2003).

Although HIV/AIDS is a major concern for all age groups, to date, limited empirical research has been conducted in South Africa on the older population. A report on poverty and older persons found that during 1999, 30% of persons 50 years and over lived in households earning less than R800 per month. A further quarter was chronically poor, meaning that they lived in households earning less than R400 per month. Africans aged 50 and over constitute almost 90% of the chronically poor (May 2003). In South Africa, among all age groups there are more females than males. As a result, 42% of all African households are female headed (May 2003). The impact of the AIDS pandemic on the population aged 50 and over is likely to be substantial given the lengthy period of illness and disability and the premature nature of the death (Knodel et al. 2001). Little has been written however about the perspective and range of experiences of older people aged 50 and above. Understanding the factors that influence the risk of HIV infection among older people may lend new insights into their perspectives and behaviour as well as how prevention programmes may be mobilised in the future to better serve their needs. In addition, the study will explore the experiences of older persons with a particular focus on their roles and activities in households affected by HIV/AIDS.

6.2 Context

Despite the relatively late arrival of AIDS in South Africa, HIV has been spreading rapidly throughout the country. However, the speed at which it is spreading varies between provinces. The province of KwaZulu-Natal has been the most severely affected by the AIDS pandemic. According to the 2007 national HIV survey, 37.4%

of women attending state antenatal clinics in the province were HIV positive (Department of Health 2008a). AIDS is the leading cause of adult mortality in KwaZulu-Natal accounting for 73% of female and 61% of male deaths at ages 15–44 years (Hosegood et al. 2004). South Africa's pandemic is growing equally fast in rural areas as it is in urban areas due to the system of migrant labour and good transport infrastructure (Whiteside and Sunter 2000). This study was conducted in an urban and rural site in KwaZulu-Natal. Both the rural and urban sites are inhabited primarily by Zulu speaking people of low socio-economic status. The study areas were purposely selected for the study to ensure representation of an urban and rural area. The rural site is KwaDumisa—a subdistrict of Umzinto-Vulamehlo District—situated approximately 80 km south west of Durban. Similar to other rural areas in South Africa many parts of the district are not readily accessible by public transport and few households have access to electricity or piped water (Aliber et al. 2005). The urban site is the former African township of Chesterville—situated approximately 15 km from the city of Durban. Chesterville, similar to other urban African townships in South Africa, is characterised by low-income formal housing development projects as well as informal settlements (Weingaertner-Kohlscheen 2003). In both areas, the level of HIV infection ranged from 27.9% to 46% (Department of Health 2008a).

6.3 Methods

The study uses a combination of qualitative and quantitative data which allows for exploration of the some of the main areas of interest from a variety of angles, and benefits from the unique insights offered by each (Simmons and Elias 1994). Ethical clearance for the study was obtained from the University of KwaZulu-Natal in South Africa.

The quantitative data was derived from a household survey. In both sites, 20 enumeration areas and 23 households in each area were randomly selected for the study. In each household, one index adult respondent aged 50 and above was randomly selected for the interview using a Kish grid. The survey questionnaire was similar for men and women. It comprised sets of questions on the following main topics: HIV-related knowledge, attitudes, and risk perceptions. In addition, a section of the questionnaire focused on illnesses and death in the household, changes in household composition, and time spent caring for children and grandchildren. For the survey, descriptive statistics were calculated first to determine the socio-demographic characteristics of the sample. A total of 974 interviews were held: 510 with women and 464 with men. The under-representation of men in our sample reflects difficulties in locating eligible male respondents in the households. For the survey, descriptive analysis was first conducted to determine the socio-demographic characteristics of the sample. The ages of respondents ranged from 50 to 97 years. Of the total sample, more than 57% were neither married nor living with a partner. At the time of the survey, 27% had no education, 31% had primary education, and 43% had secondary or higher education. The majority of the sample (60%) was not currently employed, and only 21% were in regular employment. The main source of income was some form of government grant. Almost 33% of respondents were dependent on the old age pension. Among those respondents who have ever cared for someone with HIV/AIDS, 42% had received some form of state cash transfer, while 58% received no government assistance. Among older people, more women than men were recipients of state cash transfers. To assess their own perceived risk of HIV infection, all respondents were asked the following question: "Considering all things, do you consider your chance of getting HIV to be high, medium, low, or no chance at all?" For the purpose of analysis, respondents are divided into medium–high and no- or low-risk groups. To investigate the statistical association between perception of risk of HIV infection and the major explanatory variables, the *chi-square test* was used. Logistic regression then was used to explore factors associated with a medium to high risk of HIV infection. For the purpose of the analysis, individuals with a medium to high perceived risk of HIV infection are coded 1, while those reporting little of no perceived risk of HIV infection are coded 0. The analysis also focuses on the characteristics of caregivers. For our purposes, caregivers are defined as those respondents who have ever provided care for someone who was either living with or had died of AIDS.

The qualitative data for the study was obtained from focus group discussions. The results from the focus group discussions were used to complement the information obtained from the survey. A total of eight focus group discussions were held: four in the rural area and four in the urban area. Focus group participants were recruited from several sources, including direct contact with individuals at health facilities, community based organisations, and posters placed at public venues. Each focus group discussion was made up of six to eight people. All participants were assured of confidentiality and that anonymity would be maintained at all times. In addition, they were informed that participation in the study was voluntary and they could withdraw from the study. Areas of investigation included the way in which older people perceived the HIV/AIDS pandemic, their fears regarding HIV/AIDS, and the impact of providing informal caregiving. All the focus group discussions were recorded with the permission of the participants. In addition, detailed notes were taken during the interviews. The tapes were transcribed and translated into English. All the transcripts were read and reread and organised according to particular themes and assigned initial codes. After initial coding, all the data were assembled under particular themes. In the final analysis, the codes were modified and recurrent themes that emerged across the transcripts were identified. The transcripts are used to illustrate particular findings.

6.4 Results

6.4.1 Personal Awareness of AIDS

Several studies suggest that personal knowledge of someone with HIV/AIDS may influence an individual's perception of his or her risk (Camlin and Chimbwete 2003). Personally knowing someone with HIV/AIDS may prevent denial and increase perceived risk (Camlin and Chimbwete 2003). In the survey, 58% of men

Table 6.1 Experience with illness and loss of household members and others, among all respondents

		Age		Gender of R		Location of R	
	Total	50–59	60+	Men	Women	Urban	Rural
No. of cases	974	576	398	464	510	486	488
% who personally know someone infected with HIV/AIDS	961	45.1	36.0*	45.3	37.9*	41.2	41.6
% who personally know someone who died of HIV/AIDS	956	58.7	51.8*	63.7	48.9*	60.1	51.7*
% with household member currently ill	972	18.2	27.9*	23.2	21.2	10.9	33.3*
% currently caring for ill household member	974	11.6	18.6*	12.3	16.5	6.8	22.2*
% ever experienced death in family	974	78.4	85.9*	79.7	82.9	79.5	83.4*
% experienced family death in past year	974	26.7	24.2	25.7	25.7	24.0	27.3
% with family member who died of HIV/AIDS	955	14.9	15.8	14.3	16.2	15.1	15.5
% with family member with/died of HIV/AIDS who was breadwinner	973	5.7	6.1	5.6	6.1	3.3	8.4*
% with family member living with HIV/AIDS	974	3.6	4.0	3.2	4.3	4.1	3.5
% who ever cared for someone with HIV/AIDS	971	16.9	16.4	15.3	17.9	14.0	19.4
% who ever cared for two or more HIV/AIDS cases	974	4.3	4.8	4.7	4.3	2.5	6.6*
% with at least one child in household orphaned by AIDS	974	4.3	9.1*	4.5	7.8*	3.9	8.6*
% with 2 or more children in household orphaned by AIDS	974	2.1	3.5	1.9	3.3	0.6	4.7*

*$p<0.05$

and women personally knew someone who was living and/or had died of AIDS. Older respondents aged 60 and over were less likely to report personally knowing someone who was living and/or died of AIDS than younger respondents in the age group 50–59 years. According to Table 6.1, almost 40% of respondents, more men (45.3%) than women (37.9%), personally knew someone infected with HIV/AIDS. However, a slightly higher proportion of respondents, more men (63.7%) than women (48.9%) personally knew someone who had died of HIV/AIDS. Urban respondents (60.1%) were more likely than rural respondents (51.7%) to know someone who had died of HIV/AIDS.

Overall, 39% of respondents perceived some risk of HIV infection. However, only 15% of respondents perceived a medium–high risk of HIV infection. Table 6.2 shows the relationship between respondents who perceived a medium to high risk of HIV infection by selected characteristics of respondents.

Table 6.2 Percentage of respondents reporting medium–high perception of risk of HIV infection, by selected characteristics, and odds ratios from logistic regression analysis assessing association between characteristics and medium–high perceived risk of HIV infection

Characteristic	Percent	Adjusted odds ratios
Age		
50–59	19.6*	1.00
60+	8.6	0.44*
Marital status		
Neither married nor living with a partner	14.4	1.00
Currently married or cohabiting	16.0	0.82
Gender		
Male	10.6*	1.00
Female	19.2	4.03*
Place of residence		
Urban	9.5*	1.00
Rural	20.7	3.69*
Highest level of education		
Primary or none	16.5	1.00
Secondary or higher	13.4	1.02
Currently employed		
No	14.3	1.00
Yes	16.2	0.76
State cash transfer		
No	18.7*	1.00
Yes	10.4	0.59
Ever cared for someone with HIV and AIDS		
No	12.7*	1.00
Yes	26.7	0.61
Personally know someone living with HIV/AIDS		
No	9.7*	1.00
Yes	19.0	2.77*
Number of sexual partners in the last 3 years		
Less than 2 partners	14.0	1.00
2 or more partners	24.3	2.97*
Log likelihood		–429.83
Chi-square significance		0.0000

$^*p<0.05$

According to Table 6.2, younger respondents (aged 50–59) perceived a higher risk of HIV infection than older respondents (aged 60 and above). More women than men perceived a heightened perception of risk of HIV infection, and perceived risk of HIV infection was significantly higher among rural respondents than urban respondents. Men and women with more sexual partners would be expected to have a higher perceived risk of HIV infection. The results conform to this expectation. Personally knowing someone with HIV/AIDS significantly heightened the perceived risk of HIV infection among older people. In addition, respondents who reported caring for someone with HIV/AIDS had an increased perceived risk of

HIV infection. Marital status, education, and employment status were not significantly associated with perceived risk of HIV infection.

The association between the selected characteristics and medium–high perceived risk of HIV infection was also explored using a standard multivariate logistic regression. Table 6.2 shows the odds of having a medium–high perceived risk of HIV infection by selected characteristics of respondents. In the logistic regression analysis (which included all variables), the strength of these associations were attenuated though their directions remained the same. The adjusted results show that the odds of perceiving a higher risk of HIV infection were greater among women than men. Both age and place of residence remained significant predictors of perceived risk of HIV infection. Respondents who reported two or more partners in the last 3 years were also significantly more likely to perceive themselves at a heightened risk of HIV infection than other respondents. In addition, personally knowing someone with HIV/AIDS significantly increased perceived risk of HIV infection. However, after controlling for the other variables, ever caring for someone with HIV/AIDS and having received a state cash transfer were no longer significant. The model is significant as indicated in Table 6.2.

In the focus group discussions older men and women recognise that AIDS is a widespread problem in their communities. One respondent observed, "We are burying people every Saturday and even on weekdays. The HIV/AIDS problem is all over the country" (Rural Male). Some admitted that their children had died of HIV/AIDS, while others were caring for children who were living with HIV/AIDS, as is illustrated in the following comment: "I have a daughter who is living with HIV/AIDS. She got infected in 1996 from her husband who died" (Urban Female). In addition, some older men and women are assuming responsibility for caring for their grandchildren orphaned due to the HIV/AIDS related death of one or both parents.

In the focus group discussions few older men and women expressed concern about the risk of HIV infection from their own sexual behaviour. Older people observed that younger people are more likely to be at risk of HIV infection because they have many sexual partners. However, in one focus group discussion women pointed out that older men were having sexual relations with young women and as a result older women are at risk of HIV infection. A few older men and women expressed concern about their own risk of HIV infection because of their role as informal caregivers to their sick and/or dying children. They were particularly worried about the risk of HIV infection through contact with infected bodily fluids in their role as informal caregivers.

In the focus group discussion it became clear that most older people are aware of the means to protect themselves against the risk of HIV transmission but few reported taking precautions when caring for their sick and/or dying children. They stressed that it was important to wear gloves when patients had open wounds or diarrhoea but this was rarely the case. Their precarious financial situation also means that older people do not have access to provisions such as gloves that could assist in protecting against the risk of HIV infection. This is demonstrated in the following comment:

> When your child is sick you may seek help from community health workers. However, sometimes your child messes herself and there is nobody to give you help either early in the morning, during the day or at night, nobody helps. Sometimes you need gloves but you don't get them and have to use your naked hands (Rural Female).

6.4.2 *Informal Caregiving*

Table 6.1 shows experiences with illness and loss of household members and others among all respondents. Less than one-quarter reported that a member of their household was suffering from an illness. At the time of the survey, respondents aged 60 and above were significantly more likely than respondents aged 50–59 to report that a household member was currently ill and that they were currently caring for an ill household member. Rural respondents were significantly more likely than urban respondents to report that a household member was currently ill and they were currently caring for an ill household member. Most respondents (81.4%) had experienced a death in the family. However, more than one-quarter had experienced a family death in the past year. Almost 15% of respondents had a family member who was suffering or had died of HIV/AIDS. At the time of the survey, 5.9% of respondents, more rural (8.4%) than urban (3.3%), stated that the family member who was suffering or had died of HIV/AIDS was the primary breadwinner. Only 3.8% had family members living with HIV/AIDS but 17% of respondents, more rural (19.4%) than urban (14%), had cared for someone with HIV/AIDS. Fewer respondents (4.6%), more rural (6.6%) than urban (2.5%), had cared for two or more HIV/AIDS cases. Respondents in the rural area were most likely to live in households with children orphaned by AIDS. At the time of the survey, 6.7% of respondents were living in a household that was caring for one child that had lost their parent due to HIV/AIDS. However, fewer (2.8%) were living in a household with two or more children that had lost their parent due to HIV/AIDS. Households ranged from caring for one to eight orphans, with 57% caring for one, 18% caring for two, and 25% caring for three or more orphans. The orphans were in most cases the grandchildren of the respondents.

Respondents were asked about the main caregiver to those infected and/or affected by HIV/AIDS in the household. Table 6.3 shows women (42.2%) were more likely than men (17.2%) to report that they were the main caregiver to someone infected and/or affected by HIV/AIDS. Men (6%) were more likely than women (1%) to report that they were sharing responsibility for caring for someone who was infected and/or affected by HIV/AIDS. Older people often take on the role of informal caregivers not only to their sick and/or dying children but also neighbours, close friends, and other relatives.

Women were also more likely than men to report undertaking a diverse range of caregiving activities including washing, bathing, and feeding the HIV infected person (see Table 6.4). Table 6.4 shows that women were more likely than men to report that their caregiving activities involved giving the HIV infected person their

Table 6.3 Main caregiver to household member who was infected and/or affected by HIV/AIDS by gender of caregiver

Caregiver	Men, $N=464$ (%)	Women, $N=510$ (%)
Respondent self	17.2	42.2
Respondent's partner	5.2	0.8
Respondent's son/daughter	3.0	2.0
Respondent and respondent's partner	6.0	1.0
Relative	13.2	8.0
Other	1.5	0.8
No infected or affected HH member	53.9	45.3
Total percent	100	100

Table 6.4 Percentage of respondents who reported undertaking particular caregiving activities by gender of caregiver

Caregiving activities	Men, $N=72$	Women, $N=93$
Assist with household chores	73.6	88.0*
Prepare a meal	58.3	92.4*
Feed the person	73.6	79.3
Bath the person	51.4	80.4*
Dress the person	70.8	75.0
Give medications/pills/injections	70.8	81.5
Help to the toilet	63.9	87.6*
Get them in and out of bed	61.1	73.9
Provide them with emotional support	69.4	92.4*

*$p<0.05$

medication. Not surprisingly, women were also significantly more likely than men to report that they provided emotional support to the HIV infected person.

Respondents were asked about the financial consequences on the household after experiencing a death of a family member who died of AIDS. Figure 6.1 shows that the death of the breadwinner had detrimental financial consequences for the household. Respondents were asked about the impact of the death of the breadwinner on the household. A few of the respondents reported more than one financial consequence. Almost half of respondents reported that there was an increase in household expenditure because of funeral and medical costs. Almost 9% reported financial difficulties because of an increase in medical costs while 39% reported financial difficulties because of funeral costs. In addition, 36% of respondents reported financial difficulties because of loss in income, from less labour on the farm or lower remittances or termination of grant. Some (10%) reported financial difficulties because of other reasons. Only 10% reported that the death of the person did not affect the household financially. It is likely that the majority identified the impact that they considered most important.

Table 6.5 shows the characteristics of caregivers and non-caregivers. Younger respondents (aged 50–59 years) were more likely to be caregivers than older respondents (aged 60 and above) but this was not significant. In addition, more women

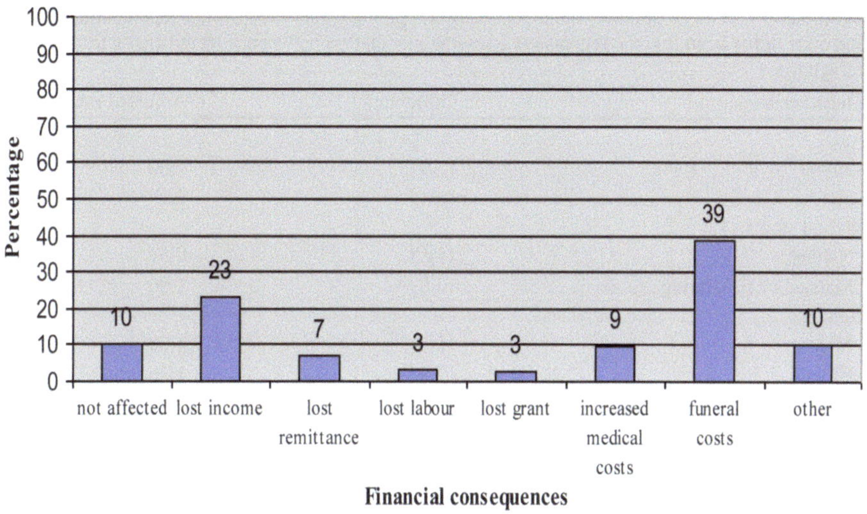

Fig. 6.1 Financial consequences for respondent's household of having a family member who died of AIDS

than men were engaged in caregiving. Caregivers were significantly more likely than non-caregivers to be neither married nor cohabiting and they were also more likely to live in rural areas. Education and employment status were not significant predictors of caregiving activities. In addition, state cash transfers did not seem to significantly influence caregiving activities. In the multivariate model using logistic regression (which includes all the variables), place of residence emerged as the dominant predictor of caregiving. Respondents in the rural area were 1.85 times more likely than respondents in the urban area to report caregiving activities. The model is statistically significant as indicated in Table 6.5.

In the focus group discussion some older people explained that they had to provide for all the basic needs of their HIV infected children. Older men and women report that they are dependant on pension funds to support not only themselves, but also other members of their household as well. Many expressed concern that the lengthy period of illness that follows HIV infection may have long term financial implications for their household. Many older people find themselves forced to spend most of their resources on caring for their sick children and, in most cases, their children seldom recover and they are not able to recuperate their losses. Some also expressed resentment that they had to provide support for their children at a time of their life when they would have instead expected support. Many expected that younger family members would provide support for them in their old age but now they find themselves having to assume primary responsibility for caring for their adult children. Some were not optimistic that their children would recover in the future. A rural woman explains some of the dilemmas that confront older people:

> HIV/AIDS affects us because they [HIV infected person] do not just get sick but they usually end up dying. Often children are an investment for their parents, as one day they will

Table 6.5 Percentage of respondents reporting caregiving by selected background characteristics, and odds ratios from logistic regression assessing association between selected characteristics and caregiving

Characteristics	Caregivers	Adjusted odds ratios
Age		
50–59	16.9	1.00
60+	16.4	0.99
Marital status		
Neither	18.8*	0.57*
Married or cohabiting	13.9	1.00
Gender		
Male	15.3	1.00
Female	17.9	1.26
Place of residence		
Urban	14.0*	1.00
Rural	19.4	1.85*
Highest level of education		
Primary or none	17.6	1.00
Secondary or higher	15.5	1.08
Currently employed		
No	17.0	1.00
Yes	16.3	0.86
State cash transfer		
No	17.1	1.00
Yes	16.1	0.83
Log likelihood		–429.83
Chi-square significance		0.0008

*$p < 0.05$

> provide help to the family when the parents get old. However, some children die leaving their children behind. Some die after their parents have sent them to school and then the parents have to start from scratch. They don't just die, it would be better if they did, instead they get sick and they finish whatever money you have by going to the doctors and buying food for them. When she dies you already have nothing as you are poor.

Some participants felt trapped in their caregiving roles because they felt a sense of obligation to fulfill their caregiving roles but they were very concerned about their worsening economic situation. Older family members often have to assume responsibility not only for the debts that are incurred from AIDS related illness but also funeral costs. When the sick patient finally succumbs to the illness older family members need money to pay for funeral costs. In addition, a few older family members also have to take responsibility for the care of grandchildren orphaned by AIDS. In the focus group discussions participants explained that as well as meeting the basic needs of their grandchildren they also had to assume responsibility for their grandchildren's health and welfare and, specifically, their education. They have to not only pay for school fees but also other indirect costs such as school uniforms and transportation to school. Many report that they use their pension to cover some of these expenses. In order to survive they are often forced to resort to other coping strategies including begging and borrowing money from neighbours and other community members. Some complain that they are not able to draw on social networks for support and often in times of adversity their friends turn their backs on them. One urban female stated:

> People will turn their backs on you and you would feel unaccepted. Sometimes if you are lucky you would have a certain source you would rely on, otherwise your neighbours would not help you with anything. And friends whom you trusted would also run away from you. What we normally do as old people who stay here in urban areas is to join groups or organisations that would help us in times of difficulty.

The consequences for their health are also severe. Many of the participants reported a number of health ailments. In most cases mothers are the main caregivers in the household and they assume responsibility for the housework. Some women expressed the feeling that their task was never-ending and they reported feeling physically and emotionally exhausted. There was a feeling that they were overworked. A few women reported that they rarely had an uninterrupted night because they were so busy taking care of their children. Sometimes even their sick and dying children cannot ignore the adverse impact on their older parent's health and well-being as is illustrated in the following comment by one urban woman: "Sometimes it is my sick daughter who feels sorry for me, saying I must also take care of myself and eat". Also, because they have limited economic resources, they are not in a position to hire someone to assist them in their caregiving activities which could have eased some of the physical burdens they face in caring for someone living with HIV/AIDS.

6.5 Discussion

Some 17% of all people in the world living with HIV can be found in South Africa (UNAIDS 2008) and the pandemic is now believed to be generalised in all sectors of the population (Shisana et al. 2005). In a number of developed countries there has been considerable focus on the impact of AIDS on the older population, however, in many African countries there is a relative paucity of empirical research conducted on the population aged 50 and over. In the past 4 decades the studies that have explored the impact of AIDS focused scant attention on older people. However, emerging evidence from a number of African countries suggests that older people are affected either directly or indirectly through their personal knowledge of and interaction with people living with HIV/AIDS (HelpAge International 2004; Nhongo 2004).The purpose of this paper is to contribute to the growing number of studies on the impact of AIDS, but focus more specifically on the older population.

The findings of the present study show that few older men and women perceived themselves at medium or high risk of HIV infection. This is in contrast to studies among younger men and women which suggest that they are aware of their risk of HIV infection (Macintyre et al. 2004; Anderson et al. 2007). A study conducted in KwaZulu-Natal found that young women living in households with a chronically ill member, most probably affected by AIDS, have a heightened risk of HIV infection

(Macintyre et al. 2004). Similarly, a longitudinal study in South Africa revealed that knowing someone who had died of AIDS was associated with an elevated perceived HIV risk for women (Anderson et al. 2007). Despite the biological fact that HIV transmission through caregiving is minimal, a number of older respondents in the present study perceived themselves at risk of HIV infection because of their personal knowledge of someone with HIV/AIDS and their role as caregivers, a finding consistent with other studies (Macintyre et al. 2004; Anderson et al. 2007). Some older people explained that they felt at risk of HIV infection because they do not have control over their situation. Their precarious financial situation in particular makes them vulnerable and as a result they are less likely to adopt risk reducing strategies to protect themselves in their role as informal caregivers. Knowledge of AIDS is an important, if not sufficient, precondition for behavioural change. Older people are often not provided with the information required for them to protect themselves against the risk of HIV infection. This is due in part to the neglect of HIV information and prevention campaigns to visibly target the older population (HelpAge International 2003). Health care workers also fall into the "trap of age related stereotypes" which can delay prevention and diagnosis. Often health care workers do not ask older patients about their sexual behaviour. In many societies, older people are often stereotyped as being non-sexual beings who should not, cannot, and do not have sex (Hall et al. 1982). It is also highly probable that some health care workers may not be fully aware of the risk of HIV infection among older people, while others may experience social barriers that prevent them from openly discussing sexual matters (UNAIDS 2004).

The AIDS pandemic has led to a reversal of roles as more and more older people are taking on the role of caregivers and providers for their chronically and terminally ill and HIV infected children and grandchildren. Historically, it was not unusual for a daughter to send her children to her mother and sometimes even to the mother of the child's father (Upton 2003). However, nowadays due to the AIDS pandemic parents are forced to assume responsibility not only of their grandchildren but also adult children. Parents often invest in the health, education, and well-being of their children with the expectation that they will care for them in their old age (Saengtienchai and Knodel 2001; Williams and Tumwekwase 2001). Increasingly, however, there has been a shift in roles and even seemingly a reversal for many as older caregivers are compelled to provide support and care for those infected with or affected by HIV/AIDS. Women were more likely than men to report that they assumed responsibility for caregiving activities. Yet men also play a significant role in providing support in families that are directly affected by the AIDS pandemic. It is clear that there is a gender division of labour in the family but men are also responding to the changing situation by performing roles that extend beyond financial support. However, several challenges were identified for both older men and women.

Given the devastating effects of the AIDS pandemic it is not surprising that almost one in five older men and women reported that they were caring for someone with HIV/AIDS. However, the number of respondents reporting that they were or have ever been involved in caregiving activities is surprisingly low and is most likely an underestimate given the high burden of HIV/AIDS in the community. It is probable that reporting biases might have influenced the underreporting of caregiving activities. Studies suggest that reporting biases occur as a result of poor understanding of survey questions, inadequate recall, and also, social desirability (Catania et al. 1990). Another potential explanation for this disparity is the fear of stigma and discrimination associated with HIV/AIDS, which might have also influenced the underreporting of caregiving activities.

In most cases, respondents reported that they were caring for a family member but sometimes also a close friend or neighbour. The study found that most caregivers are likely to be living in the rural areas rather than the urban areas. Rural caregivers may face additional challenges because of factors such as geographical isolation and lack of health facilities. When designing programmes to assist caregivers, there is a need to take these factors into consideration. Many older people explained that they felt obliged to take responsibility for the care of the HIV infected person and expressed frustration, desperation, and helplessness at their situation because of the enormous burden placed on them. Most of them did not have a high level of education and were unlikely to be currently employed. They were reliant on state cash transfers to survive and as a result they felt slightly overwhelmed by the financial responsibilities of caring for the HIV infected person. The costs of caring for an HIV infected person were ever increasing and household resources were stretched to the limit. Limited income opportunities also emerge as a major reported concern about their future. Some were also not optimistic that their sick patient was going to recover from the illness.

Older people residing in households where the breadwinner has died due to HIV/AIDS face financial difficulties, either through the loss of income or an increase in expenditure. Consistent with other studies, this study found that older people often have to shoulder the costs not only of the debts incurred from AIDS related illness, but also funeral costs (Steinberg et al. 2002). One of the most critical effects of the AIDS pandemic is that it robs the family of their only social security system; economically active members are removed from the equation when they fall ill and die, leaving children and the elderly to fend for themselves (May 2003). Social pensions such as the Old Age Pension fund and even the child grant may provide a lifeline to many older men and women directly impacted by HIV/AIDS. In South Africa the government provides the elderly with Old Age Pension, which is primarily a poverty relief programme; however, it may also contribute towards other costs such as food, clothing, medicine, and school fees. There is an urgent need for interventions to ensure greater government support to families and orphans in order to relieve their burden. Assistance is needed to help families access these grants, which currently exists but are poorly utilised due to the complexities of the application process (Madhavan 2004).

The consequences for the health of older people are also likely to be severe. As people grow older they are also more likely to experience chronic and acute illness commonly associated with ageing such as hypertension, arthritis, and diabetes, and the increased burden of caregiving may add to their health problems. Some respondents observed that they were physically exhausted and they felt completely overwhelmed with their circumstances. They found themselves undertaking a diverse range of physical activities including bathing, dressing, and feeding. In addition, in the focus group discussion it became clear that as symptoms worsen they increasingly take on clinical roles such as changing soiled linen and cleaning wounds, which exposes them to the risk of HIV infection through contact with infected blood through open cuts/sores. The situation is also compounded by their age and lack of financial resources. Other studies among older people have also reported their frustration as well as stress and emotional strain from their inability to meet the needs of their children and grandchildren (Legido-Quigley 2003).

The advent of antiretroviral treatment and the increasing availability and use of antiretroviral treatment is likely to have a huge impact on AIDS morbidity and mortality, reducing the number of sick and dying. However, there is a great degree of variability in government capacity at the district, provincial, and national levels to provide antiretroviral treatment. For a considerable period of time the government of South Africa was embroiled in a debate on the causal link between HIV and AIDS, and has displayed widespread resistance in introducing antiretroviral treatment (Maharaj and Roberts 2006). More recently, however, there has been a growing recognition of the gravity of the situation and this has led to the greater expansion of treatment in South Africa. With the increasing availability of antiretroviral treatment it is likely that the roles and responsibilities of older people will change dramatically. Effective treatment requires that patients observe strict adherence to sometimes complicated drug regimens, and older people can be a vital resource in this regard (Williams et al. 2008).

References

Aliber, M., de Swardt, C., du Toit, A., Mbhele, T., & Mthethwa, T. (2005). *Trends and policy challenges in the rural economy*. Cape Town: HSRC.

Anderson, K. G., Beutel, A. M., & Maughan-Brown, B. (2007). HIV risk perceptions and first sexual intercourse among youth in Cape Town South Africa. *International Family Planning Perspectives, 33*(3), 98–105.

Best, K. (2004). HIV/AIDS does not spare older people. *Network, 22*(1), 26–28.

Caldwell, J., & Caldwell, P. (1993). The South African fertility decline. *Population and Development Review, 19*(2), 225–262.

Camlin, C., & Chimbwete, C. (2003). Does knowing someone with AIDS affect condom use? An analysis from South Africa. *AIDS Education and Prevention, 15*(3), 231–244.

Catania, J. A., Chitwood, D. D., Gibson, D. R., & Coates, T. (1990). Methodological problems in AIDS behavioural research: influences on measurement error and participation bias in studies of sexual behaviour. *Psychological Bulletin, 108*, 339–362.

Chazan, M. (2008). Seven 'deadly' assumptions: unravelling the implications of HIV/AIDS among grandmothers in South Africa and beyond. *Aging and Society, 28*, 935–958.

Department of Health. (2008a). *National HIV and Syphilis sero-prevalence survey of women attending public antenatal clinics in South Africa—2007*. Pretoria: Department of Health.

Department of Health. (2008b). *Republic of South Africa: Progress report on declaration of commitment on HIV and AIDS. Prepared for United Nations General Assembly Special Session on HIV and AIDS*. Pretoria: Department of Health.

Hall, A., Selby, J., & Vanclay, F. M. (1982). Sexual ageism. *Australian Journal on Aging, 1*, 29–34.

HelpAge International. (2003). *HIV/AIDS and aging: A briefing paper*. London: HelpAge International.
HelpAge International. (2004). *Age and security: Summary report—how social pensions can deliver effective aid to poor older people and their families*. London: HelpAge International.
HelpAge International. (2006). *Baseline information from Africa regional HIV and AIDS project*. London: HelpAge International.
Hosegood, V., Vanneste, A. M., & Timaeus, I. (2004). Levels and causes of adult mortality in rural South Africa: the impact of AIDS. *AIDS Care, 18*, 663–671.
Joubert, J. D., & Bradshaw, D. (2006). Population ageing and health challenges in South Africa. In K. Steyn, J. Fourie, & N. Temple (Eds.), *Chronic diseases of lifestyle in South Africa: 1995–2005. Technical Report*. Cape Town: South African Medical Research Council.
Knodel, J., Van Landingham, M., Saengtienchai, C., & Im-em, W. (2001). Older people and AIDS: quantitative evidence of the impact in Thailand. *Social Science and Medicine, 52*(9), 1313–1327.
Knodel, J., Watkins, S., & Van Landingham, M. (2002). *AIDS and older persons: An international perspective*. Michigan: University of Michigan, Population Studies Center at the Institute of Social Research.
Legido-Quigley, H. (2003).The South African old age pension: Exploring the role on poverty alleviation in households affected by HIV/AIDS. Paper presented at the 4th international research conference on social security, Antwerp, Belgium. 5–7 May.
Leibbrandt, M., Woolard, I., Finn, A., & Argent, J. (2010). *Trends in South African income distribution and poverty since the Fall of Apartheid. OECD Social, Employment and Migration Working Paper, No. 101*. Paris, France: OECD.
Macintyre, K., Rutenberg, N., Brwon, L., & Karim, A. (2004). Understanding perceptions of HIV risk among adolescents in KwaZulu-Natal. *AIDS and Behaviour, 8*(3), 237–250.
Madhavan, S. (2004). Fosterage patterns in the age of AIDS: continuity and change. *Social Science and Medicine, 58*(7), 1443–1454.
Maharaj, P., & Roberts, B. (2006). Tripping Up: AIDS, Pharmaceuticals and Intellectual Property in South Africa. In C. Grown, E. Braunstein, & A. Malhotra (Eds.), *Trading women's health and rights trade liberalisation and reproductive health in developing economies*. London: Zed Books.
Mall, F. (2005). The impact of HIV/AIDS on older people. *Humanitarian Exchange Magazine, 5*, 29–31.
May, J. (2003). *Chronic poverty and older people in South Africa. CPRC Working Paper Number 25*. Manchester, UK: University of Manchester.
Moore, A. R., & Henry, D. (2005). Experiences of older informal caregivers to people living with HIV/AIDS in Lome, Togo. *Aging International, 30*(2), 147–166.
Nhongo, T. (2004). *The changing role of older people in African households and the impact of aging on African family structures*. Johannesburg: Help Age International.
Ntozi, J., & Nakayama, S. (2001). AIDS in Uganda: How has the household coped with the pandemic? In O. Orubuloye, J. C. Caldwell, & J. P. M. Ntozi (Eds.), *The continuing HIV/AID pandemic in Africa, responses and coping strategies*. Canberra: Health Transition Centre, Australian National University.
Saengtienchai, C., & Knodel, J. (2001). *Parents providing care to adult sons and daughters with HIV/AIDS in Thailand. UNAIDS best practice collection*. Geneva: UNAIDS.
Shisana, O., Rehle, T., Simbayi, L., Parker, W., Zuma, K., Bhana, A., Connolly, C., Jooste, S., & Pillay, V. (2005). *South African National HIV prevalence, HIV incidence, behaviour and communication survey, 2005*. Cape Town: Human Science Research Council.
Simmons, R., & Elias, C. (1994). The study of client-provider interactions: a review of methodological issues. *Studies in Family Planning, 25*(1), 1–16.
Ssengonzi, R. (2007). The plight of older persons as caregivers to people infected/affected by HIV/AIDS: evidence from Uganda. *Journal of Cross Cultural Gerontology, 22*, 339–353.
Statistics South Africa. (2011). *Mid-year population estimates*. Pretoria: Statistics South Africa.

Steinberg, M., Johnson, S., Schierhout, G., & Ndegwa, D. (2002). *Hitting home: How households cope with the impact of the HIV/AIDS pandemic: A survey of households affected by HIV/AIDS in South Africa*. Washington: The Henry J. Kaiser Family Foundation.

U.S. Centers for Disease Control and Prevention. (1998). AIDS among persons aged ≥50 years. United States, 1991–1996. *MMWR, 47*(2), 21–27.

UNAIDS. (2002). *AIDS and older populations*. New York: United Nations.

UNAIDS. (2004). *UNAIDS pandemic update: Fact sheet—Africa*. New York: UNAIDS.

UNAIDS. (2008). *Report on the global AIDS pandemic 2008*. Geneva: Joint United Nations Programme on HIV/AID.

UNDP. (2003). *South Africa human development report: The challenge to sustainable development: Unlocking people's creativity*. Pretoria: UNDP.

Upton, R. (2003). 'Women have no tribe' Connecting carework, gender and migration in an era of HIV/AIDS in Botswana. *Gender and Society, 17*(2), 314–322.

Weingaertner-Kohlscheen, C. (2003). *Towards integrated solid waste management in low-income housing areas in Durban, South Africa*. Stockholm: Universitetsservice.

Whiteside, A., & Sunter, C. (2000). *AIDS: The challenge for South Africa*. Johannesburg: Human and Rousseau.

Williams, N., Knodel, J., Kim, S. K., Puch, S., & Saengtienchai, C. (2008). Overlooked potential: older age parents in the era of ART. *AIDS Care, 20*(10), 1169–1176.

Williams, A., & Tumwekwase, G. (2001). Multiple impacts of the HIV/AIDS pandemic on the aged in rural Uganda. *Journal of Cross-Cultural Gerontology, 16*, 221–236.

Chapter 7
The Impact of HIV/AIDS on Older Persons in Uganda

Annet Nankwanga, Stella Neema, and Julie Phillips

7.1 Introduction

AIDS remains one of the major challenges of the twenty-first century. More than 33 million adults and children are living with AIDS worldwide, of which 22 million can be found in Sub-Saharan Africa. During 2009 alone, an estimated 1.3 million adults and children died as a result of AIDS in Sub-Saharan Africa (UNAIDS 2010). Since the beginning of the pandemic more than 15 million Africans have died from AIDS; although access to antiretroviral treatment has started to lessen the toll of AIDS, fewer than half of Africans who need treatment are receiving it (UNAIDS 2010). In addition, the number of children orphaned by AIDS alone increased drastically from 11.5 million in 2001 to 15 million in 2003; and it is estimated to have reached 24 million in 2010 (UNAIDS 2010). Over the past decade the HIV prevalence rate in Uganda has seen a dramatic decline. Estimated at about 18% in the mid and late 1980s, the HIV/AIDS prevalence rate currently stands at 6.4% (URAA 2010). This striking reduction in Uganda has been accredited to the open policy on HIV/AIDS and a strong political will to combat the pandemic. However, the impact of AIDS will remain severe for many years to come. Older people have not been spared from the pandemic that has been gripping the continent (Abrahams and Pia

A. Nankwanga (✉)
Department of Biochemistry and Sport Science,
Makerere University, Kampala, Uganda
e-mail: nankwangaa@gmail.com

S. Neema
Department of Sociology, Makerere University,
Kampala, Uganda
e-mail: sheisim@yahoo.com

J. Phillips
Physiotherapy Department, University of Western Cape,
Cape Town, South Africa
e-mail: jphillips@uwc.ac.za

P. Maharaj (ed.), *Aging and Health in Africa*, International Perspectives on Aging 4,
DOI 10.1007/978-1-4419-8357-2_7, © Springer Science+Business Media New York 2013

2002). However, in most communities worldwide, older people have not been the major focus of attention largely because of the widely held belief that they do not have much to offer since they have already played their part and have outlived their usefulness in society (Nankwanga 2011).

The impact of AIDS, especially in resource constrained settings, results in a great deal of physical and psychological suffering (Baden and Wach 1998). In addition, the high levels of morbidity and mortality associated with AIDS is likely to have long-term implications for development. Similar findings have been noted by Knodel (2008) who outlined potential pathways through which HIV/AIDS affects older persons in Thailand and Cambodia. These included emotional distress due to loss of children who are infected by the disease, difficulties in providing material support during illness, increase in caregiving activities that requires greater time and effort which eventually impacts even their involvement in income generating activities, adverse community reactions which lead to psychological trauma, stigma, and financial stress due to increased expenditure on funeral costs after death. Other impacts included loss of the child's support to the household, high costs of orphan care, and loss of future support in old age (Knodel 2008).

In many countries, the pandemic has killed middle-aged adults and shifted the burden of childrearing onto older people; this burden is growing as the number of children orphaned by the pandemic continues to increase (Ainsworth and Dayton 2000; Knodel et al. 2001). Thus, older people are now playing the key, though arduous role of bringing up children—the world's future capital. Older people also find themselves providing physical, economic, and social support to their sick children, hence having less time to engage in income generating opportunities so as to sustain their livelihoods (Tavengwa-Nhongo 2004). The pandemic has indirectly changed the role of older people, particularly women, who in Africa are less likely to have a regular income. Indeed, HelpAge International (2008a, b) highlights the gender division in caregiving activities, with 80% of older female caregivers and only 20% older male caregivers.

Most studies associate old age with persons aged 60 years and over. However, controversy still lingers on a universally acceptable definition of an older person, particularly in Uganda (Antonucci et al. 2002). For instance, the Uganda Bureau of Statistics (2007) defines an older person as any individual aged 60 years and over. This definition of an older person is also adopted in the Constitution of Uganda. However, the social protection policy being drafted in Uganda defines an older person as any individual aged 65 years and above. What then is the operational definition of an older person for this study? For our purposes, an older person is used to refer to any man or woman aged at least 55 years. Velkof and Kowal (2007) argue that it is important to take into account the lower life expectancy in the continent as over 80% of the countries in this region have a life expectancy of less than 55 years of age. In fact, life expectancy in Uganda is estimated at 53 years (UNICEF 2010), implying that the adopted age limit of 55 is still an exaggeration in statistical terms. It is however adopted based on the recommendation made by Kanyoni and Phillips (2009) after an exhaustive review of the literature in Sub-Saharan Africa.

Despite the existence of a great deal of scholarly work on HIV/AIDS, there has not been much emphasis on how the pandemic has affected older people in Uganda. This is partly because most of the data focuses on the age group 15–49 years and it is only recently that additional information on the slightly older age group of 50–59 years was collected (Kiiza-Wamala 2008). Thus, the older age groups have been excluded possibly due to the assumption that AIDS affects only young people. This assumption is, however, illusory since there is evidence that the pandemic has affected other members of society, including older people (Knodel 2008). Where scholarly attempts have been made in respect of older persons, the studies have been conducted outside Uganda (Abrahams and Pia 2002; Antonucci et al. 2002; Knodel 2008). The studies that have been conducted on older persons in Uganda (see Najjumba-Mulindwa 2004; Mugambe 2006; Baryayebwa and Barugahare 2002) have largely ignored the perspective of those affected by the AIDS pandemic. Even the studies that have focused on impacts (for e.g. Alun and Tumwekwase 2001; Bekunda et al. 2004) excluded older persons living in the urban areas of Uganda. It is not entirely clear how the disease has impacted older people and the coping strategies they employ to deal with the challenges. This paper attempts to fill this gap by exploring the impact of AIDS on older people in urban and rural areas of Uganda.

7.2 Study Context

The population of Uganda is generally young with approximately 56% below the age of 18 years and almost 92.9% below the age of 55 years. This implies that older persons as operationally defined in this chapter constitute 7.1% of the country's estimated 30 million people (UBOS 2007). Approximately 52% of the population of Uganda is female and 48% male. The older population constitutes 2.13 million in Uganda. Out of these 2.13 million older persons, 88.5% live in rural areas and only 11.5% live in urban areas (UBOS 2007). At older ages, the population of Uganda becomes more and more skewed towards women because most men were killed during past wars and insurgencies in Uganda (UBOS 2007). Older females constitute 4.2% and males 2.9% of the total population of Uganda (UBOS 2007). The older population of Uganda is growing at an annual rate of 3.2% (The Republic of Uganda 2007). At this rate, the number of older persons is likely to double in the next 25 years. Their economic conditions are harsh as official statistics show that 30% of the older population in Uganda lives in absolute poverty (Najjumba-Mulindwa 2004; Alun 2003; Lwanga-Ntale and Kimberley 2003) with limited access to employment opportunities and proper health care. The living conditions of older people are characterised by poor housing, inadequate food security, and high levels of abuse and neglect. They are socially excluded from mainstream social activities and are not protected by social safety networks that they can use to support themselves (Najjumba-Mulindwa 2004).

The study was conducted in both urban and rural sites. The sites for the study were selected from eight districts of Uganda. The districts were identified using stratified and simple random sampling. Stratified sampling was used to categorise the districts according to the four regions of Uganda, namely: Central, Northern, Eastern, and Western regions. Initially, four lists were formulated to act as sampling frames. From each list, two districts were selected using simple random sampling. The intention was to give each district an equal chance of being selected for the study. The selected districts include: Pallisa, Jinja, Mbarara, Ntungamo, Luwero, Kampala, Lira, and Nebbi. Stratified sampling was used to divide each district into a rural and an urban area. While the urban setting was considered an area within a radius of 2 km from any town in any of the selected district, the rural setting was considered as any other area situated beyond the 2 km radius. After identifying the eight districts, the researcher randomly selected four sub-counties from each district where the study was to be conducted. This was done by getting a list of all the sub-counties of each district from the District Administration Office and then randomly selecting the sub-counties from each district. Older persons were purposively sampled from each of the sub-counties. After selecting the sub-counties, guidance was sought from sub-county leaders regarding the most effective strategy to access older people. These leaders helped to direct the researchers to lower community leaders who were then requested to help identify older people in their jurisdictions. In each of the selected sub-countries, four to seven older people were identified. A higher number of older persons were selected from the Kampala district because they were easier to access in this district than in any other district. The higher number was intended to generate data that was needed to reach the point of saturation during data collection. In order to be eligible for the study men and women had to be aged 55 and over and, also, they had to be willing to share their experiences with the researcher.

7.3 Methods

The qualitative methods used in the study included a combination of in-depth interviews and focus group discussions. In total, 165 older persons were selected for the study, with 64 respondents selected from urban areas and 101 from rural areas. The number of older people interviewed was relatively large for a qualitative sample but it was necessary for purposes of covering all the regions and targeted districts and sub-counties of Uganda. The majority (94%) of the older people were over 60 years of age. In addition, more women (54%) than men (46%) participated in the study. Each interview session was conducted in a highly informal, conversational, and face-to-face style at the respondent's home. The researcher would first ask the older person whether to use English or the local language. If the older person agreed to be interviewed in English, the researcher would then proceed with the interview. If the selected older person was not in a position to answer the questions in English, the principal researcher would switch to the local language. Each interview lasted approximately one and a half hour. Following the individual interviews, focus group discussions were held. Focus group discussions were conducted to complement the information received from the individual interviews. In total, four focus group discussions were held. Of these, two were held with older persons selected from urban settings and the other two were carried out with older persons selected from rural settings. The interviews were held in convenient places such as the community leaders' offices, but when a room was not available the interviews took place in a private, quiet place under the trees. Each focus group session lasted between one and a half and two hours.

The topics that were discussed included, among other issues: main problems affecting older persons, how these problems are dealt with at the community level, problems experienced with accessibility to resources and services, the impact of HIV/AIDS on older persons, the role of parenting by older persons and its challenges, and possible strategies that could help improve their livelihoods. All respondents were given the assurance that their responses would be kept confidential and that they were free to withdraw from the study at any time if they so wished. No identifying names were used for purposes of ensuring the anonymity of the respondents. In addition, permission for using a tape recorder during the interview was obtained from each respondent. Ethical approval for the study was granted from the University of the Western Cape, South Africa, and the Uganda National Research Council of Science and Technology.

7.4 Data Analysis

The analysis involved reading the written scripts, one by one, and in a repeated manner so as to develop meaning out of them. There was also repeated listening and re-listening to recorded audiotapes, so as to reflect on the data and also make notes and memos. This was followed by grouping of the data into themes and then sorting the themes according to broad categories. This was done by combining related themes and renaming them as one theme. The interpretative technique was used to develop the categories. The developed thematic categories were then re-categorised. Sub-themes were also developed under each of the major themes.

Information under each theme was then coded into a number of smaller categories or sub-themes. This was followed by identification of overlapping codes which were placed into segments while the un-coded text that was not relevant to the research objectives was left out. In addition, field notes were used to corroborate the themes and to assist in the interpretation of findings, and where possible thick descriptions using actual words of the respondents were used for verification.

The processes were repeated continuously to make necessary refinements to the categories and are also used in the selection of appropriate quotes that convey the real meaning of the categories. It is vital to note that some of the responses from the interviews and focus groups were incorporated verbatim in the text of the study with minor editing, where need arose. They were incorporated according to the different categories of themes in order to avoid losing the original meaning.

7.5 Results

7.5.1 *Impact of AIDS*

The effects of HIV/AIDS on older people in Uganda were established by asking them to describe how the pandemic had affected them and their families. Table 7.1 presents the various effects of HIV/AIDS on older people in Uganda. While a quick glance at the results suggests that the disease affected the largest proportion (27.9%) of these people by leaving them with a burden of orphans, a more careful analysis reveals that the combination of those who lost all or some of their children, spouses, and grandchildren to the pandemic constituted the majority (58.1%). This indicates that the most horrendous effect of the pandemic constituted death of close kin, especially children and spouses. Table 7.1 indicates further, while the death of children was felt more by urban-based older people (60.6%), the burden of children orphaned by the disease was felt more by the rural-based older people (38.6%). This indicates that the effect of the disease varied across settings.

Older men and women are also at risk of HIV infection. A few reported that they were living with the virus, with a higher proportion in the rural areas than the urban areas. Interestingly, men and women in both sites felt that the disease has humbled them and increased their faith in God. The fact that HIV/AIDS caused enormous effects on Uganda's older people in terms of killing their children and spouses as well as leaving them with the burden of orphans is shown in Box 7.1.

Table 7.1 Impact of HIV/AIDS on the older people in Uganda

	Rural		Urban		Total	
	N	%	*N*	%	*N*	%
Led to death of all my children	9	14.1	4	4.0	13	7.9
Has killed some of my children	15	23.4	5	5.0	20	12.1
Has left me with a burden of orphans	7	10.9	39	38.6	46	27.9
Claimed the life of spouse	6	9.4	17	16.8	23	13.9
Killed my children and is now killing grandchildren	24	37.5	16	15.8	40	24.2
I am infected with it	2	3.1	11	10.9	13	7.9
Has brought me many sicknesses and weakened me	1	1.6	6	5.9	7	4.2
Has humbled me and increased my faith in God	0.0	0.0	3	3.0	3	1.8
Total	64	100.0	101	100.0	165	100.0

Box 7.1 Verbatim Responses on the Impact of HIV/AIDS

AIDS has greatly affected my family; my four children were laid to rest in the banana plantation over there. They were all boys (Female, aged 70, Mbarara).

I have orphaned grandchildren whose parents died from AIDS. One was taken away and I have remained with one who now has dropped out of school in senior four due to lack of money to support her (Female, aged 60, Kampala).

AIDS has greatly affected my home because the father of my children, plus some children died. I am also infected and I am on antiretrovirals. I still have to work for my children, educate and feed them alone. All my family members rejected me because they said that I am the one who killed their son with the disease, so I am struggling to bring up my children alone. I cannot even manage to get drugs to treat myself because the money I get is very little to cater for all our needs (Female, aged 61, Pallisa).

AIDS has affected my family because I lost one child and I have orphans. Because of AIDS, my grandchildren and I are suffering. I cannot provide support. I used to stay with them but when I could not manage to provide food, relatives from their father's family took them away (Female, aged 60, Luwero).

My children got AIDS and many of them and their mother died. Out of the seven children I had, only one is remaining (Male, aged 72, Nebbi).

AIDS has affected me because it killed most of my brothers and sisters. They left a number of orphans that I have the responsibility to look after. Now even some of my own children are also infected although they haven't died (Male, aged 60, Jinja).

The thematic results indicate that HIV/AIDS has caused many older people in Uganda to lose their children and to shoulder the burden of looking after orphans. Some older people face discrimination and stigmatisation as a result of suffering from the disease. Indeed, the results show that one of the respondents was abandoned together with her children by the relatives of her late husband, accusing her of having brought the disease which killed their son. Such casting of blame on the women was not uncommon in this study setting. The in-depth interviews alluded to the fact that stigma was still prevalent and those who were HIV positive were seen as worthless with nothing important to add to the community. Such an attitude is likely to negatively impact older people who are HIV positive. Focus group discussions suggest that AIDS has impoverished the households of older people as they were unable to continue with any work that would provide some income as they were intermittently ill. Some of the resources they had were used for caring for themselves including the orphaned children in their household. In addition, the death of their spouses impacted negatively on women particularly because they were often not

Box 7.2 Verbatim Responses the Impact on Widows

When I lost my husband I didn't know that he had died of HIV; I thought of different things altogether. But I came to know when my child and I fell sick, I went to hospital where they checked us and told me that both I and my child were infected with HIV... I was not working so I felt stressed economically. I could not afford to maintain the home alone because I was poor and had no job. I was being helped by neighbours who collected money for me and I started a business of selling charcoal (Female, aged 63, Pallisa).

When I lost my husband I had three children, we were in a rented house of one room. My two children were okay but the third child was infected. I came to know this when he fell sick and I took him to hospital where he was tested and found HIV positive. I treated him for sometime but he later died. We were very poor because he was a builder but when he fell sick he could no longer continue to build. I was a teacher but the small income I was getting was not enough. When he died he left me the burden of looking after orphans and educating them but I totally cannot manage the load (Female, aged 60, Jinja).

I lost my husband when I had not known that we were infected. We had four children and another three from outside. I even did not know them until the time when we were burying their father and that is when they introduced them as children of my late husband. The clan people treated me badly. I had to leave with my children and start a new life as a widow and struggle alone to raise the children (Female, aged 65, Kampala).

working and dependent on them and, as a result, they found themselves destitute after the death of their husbands. Some women reported that they were abandoned by family members and they had to fend for themselves. This in a way caused them to remain in abject poverty as illustrated by the quotes in Box 7.2.

It is clear that there is inadequate support provided to older people, if any is given at all. Governments do not usually provide economic support or subsidies for older people forced to care for sick or orphaned children. As a result, the AIDS pandemic has placed economic pressure on older people. The interviews revealed a chain of factors such as the decline in health status, high cost of medical fees during illness, loss of remittances from sick children, and multiple caregiving activities at home, which have compromised older people's ability to earn an income so as to sustain their household. Thus many older people have resorted to selling of assets to pay for expenses such as school fees, medical bills, feeding, etc. When older people were asked whether they had income generating activities that could help them earn a living, many of them replied negatively reporting lack of time, capital, and energy to carry out such activities. The few who replied affirmatively were involved in road side selling of small items. Unfortunately observations revealed that the stock was very small consisting of very few items that could not even raise a total of US$3 as capital. Their meagre earnings could not guarantee a meaningful livelihood for older people caring for sick and/or dying children or grandchildren. A comparative analysis of the responses by setting revealed that older persons who carried out such economic activities were mostly from urban settings in the central, western, and eastern regions of Uganda. There were virtually no older people carrying out such activities in northern Uganda because of the long civil war that was going on in that part of the country at that time.

However, while there were adverse economic effects for both the rural and urban areas of the country, the situation was even worse in the rural areas because of the extensive poverty and secondly because of the lack of government social protection mechanisms in place. Like many other developing countries in Africa, there are no government health insurance schemes that could cover their medical expenses. Thus, the fact that many older persons indicated that they did not have income generating activities suggests that many older people in Uganda suffered from severe economic consequences.

Many older men and women reported that they were feeling stressed because they still considered AIDS to be a death sentence. As a result, they had very little hope for the future. Some were worried about the future of their children and grandchildren. They were feeling somewhat overwhelmed with the additional responsibilities that they were forced to assume. There were some concerns about loneliness. Older men and women admitted that some members of the community were reluctant to associate with them. As a result, some decided not to reveal their status to family members and friends. Many also expressed feelings of depression and felt that they could not change their situation. Box 7.3 highlights some of the social and psychological impacts of AIDS on older people.

Box 7.3 The Social and Psychological Impact of AIDS

You feel psychologically tortured. Sometimes some people go mad. You are not happy ever because you know any time you are going to die (Male, aged 70, Ntungamo).

Sometimes you do not make developments because you think you are going to die anytime and leave your property for people (Male, aged 67, Pallisa).

You feel so bad; sometimes you remain lonely with no friends because most people may not want to associate with you (Female, aged 64, Nebbi).

Looking after orphans is like starting life all over again; I have to work for them, feed them, and provide clothes for them. I am not sure if I have the energy to cope with this situation (Female, aged 65, Jinja).

(continued)

Box 7.3 (continued)

I had a fear to tell any of my friends, sisters and other relatives, I had to keep it a secret to myself. Then when I went to hospital, they told me to bring a person to sign for me the papers for treatment that is when I had to open up to my sister. And it is my sister who revealed to other relatives which made me feel bad. But later on I came to terms with it. Up to now I feel stigmatised when people talk about HIV in my presence. I will think they are pointing a finger at me (Female, aged 60, Jinja).

My children and I are infected. I normally think of how my two children will remain when I die. Who is going to treat them? The only prayer is that God should keep me alive so that my children can first grow then I can die because there is no one who can help them when I die (Female, aged 62, Kampala).

HIV affected me socially because I became so small among my friends and relatives. I was not free as I used to be. I grew thin and felt I was not myself and I was alone because my husband abandoned me. I could not fit in any other group. I used to feel so small thinking that if my friends came to know about it they will abandon me too (Female, aged 65, Kampala).

First I thought of killing myself then I thought of running away from the family but unfortunately I had children whom I thought would be left to suffer. It was not easy for me. I took some time to settle down I started growing thin because of thoughts; I could not eat, and even talk to any relatives all I wanted was to stay alone in my own world (Female, aged 60, Mbarara).

Some older men and women who disclosed their status felt that they were blamed for their behaviour. Many members of society still hold the commonly held misconception that older people should not engage in sexual activities. This attitude has prevented older men and women from accessing health services. Older men and women who are HIV positive are sometimes shunned by their family. They are forced to leave the house and find alternative accommodation and therefore find it difficult to support themselves. Some of the respondents expressed rejection by the relatives as a result of being infected by HIV, as is illustrated by one comment: “All the relatives ran away from me and no one would give me help after having known that I was helpless and infected since HIV was basically associated with prostitution” (Female, aged 70, Lira). Another respondent observed, “I had to forget about all my relatives and live alone with my children because no one wanted to associate with me or give me help. So my life depended entirely on God who created me” (Female, aged 68, Kampala).

7.5.2 Coping Strategies

When asked how they survived without any income generating activities, a few older people from the urban areas indicated that they were dependent on family and friends. They indicated that much of the support was provided by their children. Their children often provided them with food and medical supplies. A few older people from the rural areas also indicated that they received support from their children mainly in the form of food, clothing, and money. This is indeed worrying given that more and more adult children are succumbing to the AIDS pandemic.

Older people were asked to mention any organisation that provides them with any form of support. Only a handful of older persons were able to identify organisations, which included *Munnomukabi* (A Friend in Need is a Friend Indeed) established at Makindye in Kampala district in central Uganda, *Zarimpeke,* an association for older women in Kawempe II Zone, the AIDS Association of Kampala City Council, *Agalyawamu Gegaluma Enyamma* (Unity is Strength) which was also established at Makindye, Kampala district in central Uganda, and the Uganda Reach the Aged Association (URAA) located in central Uganda. However, the majority of the older people reported that they were not receiving economic assistance from these organisations. These organisations do not reach a large sector of the population but tend to work with limited communities in their area.

Some organisations were assisting older men and women. Older people felt that these organisations were making a clear difference in the lives of the elderly and they have an important role to play in assisting the community in income generating activities. This is clearly illustrated in the following comment by one respondent:

> I was born in 1958, tested for HIV/AIDS in 1995 and found that I was positive. My husband died in 1995 and I am now a widow. I fell sick and got skin rashes, vomiting and diarrhea. I started getting treatment and I am now better. I suffered emotionally; I was being referred to as a person who did not have any use in the world. I felt hungry as there was no food. Then I came across this organisation called URAA which taught me prevention, treatment and nutrition and how to live positively. They also trained me to help others who are living with HIV. And how to do income generating activities and gave me some money to start. Now I live very well and I am okay. I thank URAA for all the help they gave me' (Female, aged 65, Kasese).

Another also stated: "...you would not have seen me here if it was not for this NGO because I can now eat and drink because of URAA. I also look after so many orphans" (Male, aged 70, Kasese). It is clear from these statements that support from organisations can help improve the life of older persons, but unfortunately there are only a handful of such organisations in the country. In addition, they are few older people who are able to access such opportunities since the majority of them are based in the rural areas.

Older people used a number of strategies to cope with the impact of the AIDS pandemic, as shown in Box 7.4. Some were concerned about the immediate impact

Box 7.4 Mechanisms Used to Cope with HIV/AIDS

Ever since I heard of and experienced the wrath of HIV/AIDS, I started emphasising to my children and grandchildren to be extra careful. I take the time to talk to my remaining children about AIDS. I advise the grandchildren to be careful when choosing partners and not to play around with girls or boys, or to use condoms just in case… Although I get little money through selling food, I ensure that I provide for them educationally for their future self-sustenance while I am still with them (Female, aged 62, Kampala).

I always tell my grandchildren to use condoms when they find themselves in the situation that can lead to their getting infected by HIV/AIDS (Male, aged 72, Luwero).

I sell cooked food by the road side to get some money which I use to feed my family and also pay school fees for my orphans (Female, aged 70, Jinja).

on their household and therefore were growing their own food to feed their children. Others approached their children for assistance or other members of their community. A number of older men and women also started antiretroviral treatment. Some joined support groups to help them cope while others turned to religion. The AIDS pandemic also encouraged older people to talk to their grandchildren about protecting themselves against the risk of AIDS.

Older men and women were asked how they wanted government to help address the impacts of this pandemic. While a large proportion of older people (17.2%) recommended that government should initiate a special fund for HIV/AIDS orphans and provide free antiretroviral to HIV positive older people (17.2%), others suggested provision of food (17.9%) and sensitisation of grandchildren about HIV/AIDS, and providing them with free condoms (16.7%). Despite their considerable caregiving activities, older people remain largely hidden from the international HIV/AIDS agenda. This is because it is assumed that they are not at risk of HIV infection, and hence receive minimal attention in HIV prevention programmes. They often lack basic information on prevention and transmission yet they provide support for their terminally ill children and grandchildren. In addition, they are also sexually active yet ignorant about prevention mechanisms which increase their risk of HIV infection. They need more support in their caregiving role. Resources and information provided to them may enhance the efforts they make to care for their children and orphans and their own self-protection. Education campaigns are also likely to alleviate fears about casual transmission, thus contributing to lower levels of stigma. Some of the suggestions by the older people are also summarised in their own words in Box 7.5.

Box 7.5 Strategies for Addressing the Effects of HIV/AIDS

What has increased the prevalence of HIV/AIDS thus far is that people are not educated about it. They are not careful about who they go out with. Long ago people used to investigate the background of the person before getting involved with them. But now parents and the government have left their roles of keeping the morals, giving children the freedom of making their own choices and decisions while still in the schools. Most of the children are taken unaware because they don't know what to do when their sexual desires are high. They behave wildly. Government should sensitise the youth about the dangers of contracting HIV/AIDS. It should provide free education to HIV/AIDS orphans up to completion. Universal primary education is not enough. The government should ensure that people are always educated about HIV/AIDS, in schools, in villages, communities, and nationally because people are still ignorant about this pandemic (Male, aged 85, Lira).

I think that poverty has been so instrumental in sustaining the impacts of HIV/AIDS. Many girls go for men with money. The government should fight poverty. Let government provide the elderly looking after orphans with financial help. Government should give a hand to the elderly persons looking after grandchildren whose parents died of AIDS. I call upon the government to set up a special fund to help HIV/AIDS orphans in educating and feeding them. Government should also support non-governmental organisations that look after orphans. It is through the non-government organisations that the government can provide school fees and medical care to the orphans. The government should distribute free condoms in villages, public areas and in all accessible areas for the young ones to use (Male, aged 72, Kawempe).

The government should ensure that drugs (ARVs) and VCT services are provided free for those infected with the virus. Government should ensure that drugs are readily available in hospitals and health centres. Government should also provide HIV positive elderly people with foodstuff (Female, aged 70, Jinja).

7.6 Discussion

The findings suggest that a number of older men and women have lost their children to the AIDS pandemic. By killing their children, HIV/AIDS has created a huge burden for older people. The findings indicate firstly that the loss of a child is a big blow not only because most parents naturally find it difficult to come to terms with it, but also because they have no one to turn to for support and care when they enter old age. Losing children thus comes with adverse effects on the survival of older persons, especially in Africa where, according to Kollapan (2008), most of them depend on their children. By killing some or all of their children, HIV/AIDS has effectively rendered older people in Uganda hopeless. They have no one to turn to for care and support. Secondly, HIV/AIDS alters family structures such that when adult children die older people once again become parents for the second time, providing care for orphaned children. Caring responsibilities have financial implications that include meeting the costs of daily living, providing food, treatment, shelter, and clothes as well as funding their education and scholastic materials. However, older people often get distressed by their failure to provide for the multiple needs of the children. As a result many of the children under their care end up not going to school, hence developing bad habits and behaviours, and eventually they become a burden to the entire community.

Evidence shows that the number of grandparents caring for AIDS orphans in developing countries has doubled over the last ten years and up to half of the world's 15 million AIDS orphans are being cared for by a grandparent (UNAIDS 2010). The majority of older caregivers are women who face serious financial, physical, and emotional stress due to their belated caregiving responsibilities (Kawuma 2011). In Uganda, the AIDS pandemic has created an enormous burden of orphans, who are most often cared for by older people. The thought of having to look after a large number of orphans when the older persons cannot even fend for themselves is not only overwhelming but also humbling at the same time. In the interviews many older people stated that they felt humbled and their faith in God increased. This situation suggests high levels of helplessness and despair, which, if not urgently addressed, can easily plunge the affected older persons into trauma, depression or other forms of extreme psychological stress. Older people indeed are faced with a number of challenges. They face problems of exhaustion as they look after sick children; in addition, they have to face grief and stress. The strain they face while caring for their children living with HIV/AIDS in the last stages of the illness can be harrowing and tiring and as a result it takes its toll on the older person's health.

Older people are also sometimes themselves living with the virus (Best 2002). Since culturally AIDS is associated with sex, prostitution or death, and older people are expected by society to have stopped engaging in sexual relationships, thus when they are infected with HIV it stigmatises them. For this reason, Atekyereza and Kirumira (2004) advised that controlling stigma is very critical in controlling the health impacts of HIV/AIDS. This implies that if the trauma felt by the infected older persons is not addressed through administering proper treatment mechanisms, they are likely to get psychologically worse. There is therefore a need to develop health care programmes that give older persons priority when they go for treatment or introduce age sensitive programmes to embrace the needs of older people. And indeed it is clear that there are few AIDS programmes that directly target older people, though they are considered to be at risk of contracting HIV. This could probably be emanating from the limited available data regarding HIV prevalence in older persons since most collected data is only for adults aged 15–49 years, as argued by Kiiza-Wamala (2008). On the other hand, it could also be emerging from the false assumption that older persons are not actively engaged in sexual relationships.

The results indicate that Uganda's older people have developed a number of mechanisms to deal with the impact of HIV/AIDS. The findings are therefore largely in line with the observation made by Apt (1996) that coping mechanisms used by the older persons tend to span over a wide spectrum stretching from initiatives and attempts made to overcome barriers confronted at personal, household, and community levels, and in the form of constraints to health care, food security and nutrition, housing and accommodation, and other forms of limitations. A close examination of the mechanisms reveals that although older people make attempts to overcome the constraints encountered in their efforts to earn a living, the attempts are not comprehensive enough. These mechanisms help the older persons to mobilise economic support and food that they need to feed the children and themselves. It is crucial to point out that although growing food may help to overcome such constraints, the extent to which it does so tends to decline as food growers grow old amidst growing numbers of people for whom the food is grown. Accordingly, the efforts that the older people in Uganda spend in growing the food are not enough when viewed from the perspective of the large numbers of grandchildren that they look after. The efforts therefore need to be enhanced by government as a matter of fulfilling its obligation to the people.

Specific emphasis needs to be placed on helping these older people by supporting the large number of orphans left in their care. While there have been attempts to set up youth-friendly services, more attention needs to be focused on developing services that meet the needs of older people. Even the community education channels, such as mass media, community events, bars, dances, music, drama, and antenatal clinics that are used to sensitise the public about these services, leave out older persons (HelpAge International 2006). Even when a poster is made, most indicate images of young people which reveals a picture that HIV/ AIDS is a young people's disease. However, increasingly, older people are infected and affected by AIDS.

7.7 Conclusion

Generally the older people in the Uganda study suggest that they are heavily affected by the AIDS pandemic. Policies and programmes should therefore focus not only on younger people, as is largely the case in Uganda, but should include and even give priority to older persons, especially in light of the fact that in addition to being infected by HIV/AIDS, the pandemic brings worse effects on the older persons than on younger people in terms of leaving some of them childless and creating a huge burden of orphaned grandchildren which becomes their responsibility. The impact of AIDS is linked to many other problems, such as poverty, inadequate nutrition, and poor public infrastructure. Efforts to fight the pandemic must take these realities into account.

References

Abrahams, D., & Pia, M. (2002). *AIDS and trends in food production in Sub Saharan Africa*. Rome: FAO.

Ainsworth, M., & Dayton, J. (2000). *Is the AIDS pandemic having an impact on the coping behaviour and health status of the elderly? Evidence from Northwestern Tanzania*. Geneva: USAID.

Alun, W. (2003). *Ageing and poverty in Africa. Ugandan livelihoods in a time of HIV/AIDS*. Hampshire: Ashgate Publishing Company.

Alun, W., & Tumwekwase, G. (2001). Multiple impacts of the HIV/AIDS pandemic on the aged in the rural Uganda. *Journal of Cross-Cultural Gerontology, 16*(3), 221–236.

Antonucci, T. C., Okorodudu, C., & Akiyama, H. (2002). Well-being among older adults on different continents. *Journal of Social Issues, 58*(4), 617–626.

Apt, N. A. (1996). *Coping with old age in a changing Africa*. Aldershot, UK: Avebury.

Atekyereza, P. R., & Kirumira, E. K. (2004). The impact of the AIDS pandemic on families and family coping strategies in Uganda. *Research Review, 15*, 33–44.

Baden, S., & Wach, H. (1998). *Gender, HIV/AIDS transmission and impacts: A review of issues and evidence*. Brighton, UK: Institute of Development Studies.

Baryayebwa, H., & Barugahare, W. (2002). *Health needs of older persons: An assessment report on Kamuli and Wakiso districts*. Kampala, Uganda: Government of Uganda.

Bekunda, R., Kibaalya, W., Rwibasira, J., Asaba, S., Haag, A., Camilo, M., et al. (2004). The burden of HIV/AIDS on the elderly in a rural setting: A case of AFXB Luweero-Uganda. International Conference on AIDS. 11–16 July.

Best, K. (2002). HIV/AIDS does not spare older people. *Network, 22*(1), 26–28.

HelpAge International. (2006). *AIDS: The frontline. Supporting older people living with HIV/AIDS and orphaned children in Mozambique, South Africa and Sudan*. London: HelpAge International.

HelpAge International. (2008a). *Ageing issues in Africa: A summary*. London: HelpAge International.

HelpAge International. (2008b). An untapped resource: How supporting older people with social protection will help achieve the millennium development goals. London: HelpAge International. http://www.chronicpoverty.org/pdfs/2003conferencepapers/Mulwinda.pdf. Accessed 25 Apr 2007.

Kanyoni, M., & Phillips, J. (2009). Factors associated with physical activity levels among older adults in selected institutions in Rwanda. *Journal of Community and Health Sciences, 4*(1), 8–14.

Kawuma, E. (2011). Impact of HIV/AIDS on older persons—current evidence. Paper presented at the National consultative meeting on mainstreaming ageing into HIV/AIDS Responses. Kampala, Uganda. 27–29 Apr 2011.

Kiiza-Wamala, J. (2008). The needs of older carers. Paper presented at the National consultative meeting on older carers of OVCs/PLWA. Kampala, Uganda. 19 November.

Knodel, J. E. (2008). Poverty and the impact of AIDS on older persons: evidence from Cambodia and Thailand. *Economic Development and Cultural Change, 56*(2), 441–475.

Knodel, J., Saengtienchai, C., Im-em, W., & Van Landingham, M. (2001). The impact of AIDS on parents and families in Thailand: a key informant approach. *Research on Aging, 23*(6), 633–670.

Kollapan, J. (2008). The rights of older people—African perspective. http://www.globalaging.org/elderrights/world/2008/africa.pdf. Accessed 13 Oct 2011.

Lwanga-Ntale, C., & Kimberley, M. (2003). *The face of chronic poverty in Uganda as seen by the poor themselves*. Manchester, UK: Chronic Poverty Research Centre.

Mugambe, B. (2006). *Targeting and protecting the chronically poor in Uganda: A case for the elderly. Policy Brief No. 3*. Manchester, UK: Chronic Poverty Research Centre in Uganda.

Najjumba-Mulindwa, I. (2004). Chronic poverty among the elderly in Uganda: Perceptions, experiences and policy issues. Institute of Statistics and Applied Economics, Makerere University. http://cprc.abrc.co.uk/pubfiles/Mulinda.pdf. Accessed 28 May 2012.

Nankwanga, A. (2011). *Exploration of needs, problems and living experiences of older persons in Uganda: Implications for policy and decision makers*. South Africa: University of Western Cape.

Tavengwa-Nhongo, M. (2004). Impact of HIV/AIDS on generational roles and intergenerational relationships. Paper presented at the National Council Workshop on HIV/AIDS and Family Well-being. Windhoek, Namibia, 28–30 January.

The Republic of Uganda. (2007). State of Uganda population report 2007. Planned urbanization for Uganda's growing population. http://www.popsec.org/documents/State%20of%20Uganda%20Population%20Report%202007.pdf . Accessed 17 Oct 2011.

Uganda Bureau of Statistic (UBOS). (2007). *Uganda demographic and health survey 2006*. Kampala, Uganda: UBOS.

Uganda Reach the Aged Association. (2010). *Advocacy and the older persons in Uganda*. Kampala, Uganda: URAA.

UNAIDS. (2010). *Report on the global AIDS pandemic*. Geneva: UNAIDS.

UNICEF. (2010). Uganda. Statistics. http://www.unicef.org/infobycountry/uganda_statistics.html. Accessed 17 Oct 2011.

Velkof, A. V., & Kowal, P. R. (2007). *Population Ageing in Sub-Saharan Africa: Demographic dimension 2006*. Washington, DC: US Government Printing Office.

Chapter 8
On the Fringe of Poverty: Care Arrangements for Older People in Rural Cameroon

Charles Che Fonchingong

8.1 Introduction

Providing care for older people living in rural areas is increasingly problematic warranting a redefinition of policy directions. Demographic shifts point to a significant increase in rural elderly with different survival challenges (Mba 2010). Very little is known about the rural elderly in Cameroon, and more concerning is the preparedness of the country to meet the challenges of its growing older population. With a population of approximately 18,060,382 inhabitants, there are almost 917,520 persons aged 60 years and over cutting across ten administrative regions. Approximately 43% of Cameroon's populations are resident in rural areas (IFAD 2008; World Bank 2009). With an ever increasing ageing population resident in rural areas, any more tinkering by Cameroon's government may prove costly. This chapter aims to make a modest contribution by uncovering support mechanisms and chronicling social realities of rural elderly. It presents variations and differential categorisations amongst rural elderly, unearthing coping strategies to stave off poverty by drawing on social capital theory. Social capital represents the whole stock and mix of social, economic, and cultural resources mobilised through individual and collective efforts which yields tangible and intangible benefits (Fonchingong 2005).

Almost 42% of Cameroon's population are below the age of 15 years with males estimated at 3,443,505 and females at 3,367,571. The majority of the population is concentrated in the 15–64 age category, which constitutes 54.5% of the population with males estimated at 4,431,524 and females at 4,392,155. The population aged 65 years and over stands at 3.4% with males constituting 253,242 and females 296,751 (United Nations 2007a, b). Two points can be gleaned from the data: firstly, the large concentration of the population in the 15–64 age category suggests a likely surge in the elderly

C.C. Fonchingong (✉)
Canterbury Christ Church University, Department of Health, Wellbeing and Family, CCCU, Canterbury, UK
e-mail: che.fonchingong@canterbury.ac.uk

P. Maharaj (ed.), *Aging and Health in Africa*, International Perspectives on Aging 4,
DOI 10.1007/978-1-4419-8357-2_8, © Springer Science+Business Media New York 2013

population, and secondly, the gender differentiation suggests a growing feminisation of the older population. Worryingly, most estimates suggests that only 10% of citizens are covered through formal social security programmes (mostly pensions), leaving 90% uncovered, with the vast majority residing in rural areas. The rural elderly are in a precarious situation, their vulnerability exacerbated by the dearth of basic infrastructure and amenities like water, electricity, health centres, and roads. As is the case in most of Africa, functioning social security programmes are in short supply and most governments struggle to cater for its ageing population (Bailey and Turner 2002; van Ginneken 2003; ILO 2001). Viable social protection programmes can no longer be delayed given the emerging demographic transition and significant shifts with implications for development policy (Kalasa 2001; Darkwa and Mazibuko 2002; Mba 2010).

Poverty is a major bone of contention and cause for concern in Cameroon with variations across regions (Abwa and Fonchingong 2004). The first national household survey as far back as 1996 estimated that 51% of the population was living in poverty. The figure had fallen to 40% by the time the second survey was conducted in 2001. However, the decline mainly benefitted urban dwellers. As a corollary to the 1996 survey, the IFAD 2008 survey indicates that 22% of the people in urban areas are poor compared with nearly 50% in rural areas. Given substantial regional variations in income poverty across agro-ecological zones in Cameroon, a study on living standards and poverty (Ministry of Economy and Finance 2002) noted that capital cities like Yaoundé and Douala had poverty rates of 7.9% and 8.6%, respectively. In other cities the poverty rate is 7.8%; rural forest regions stand at 29%, rural high plateau at 33.6%, and rural savannah regions at 40.6%. The study puts urban poverty at 13.6% and rural poverty at 34.7%. Apart from insufficient disposable income, location, access to water, health and sanitation, roads, housing, gender, and level of education are other major indicators of non-monetary poverty in Cameroon.

Although data may be obsolete, extrapolations point to rural areas being the hardest hit. Older people in rural areas are worst-off, their plight compounded by an increase in material poverty, outmigration of younger populations and rising unemployment, which poses a major threat to erstwhile traditional care arrangements. A report by the United Nations (2007a, b) makes a compelling point: a high proportion of rural communities and especially rural poor are directly or indirectly dependent on agriculture through farming, food processing, fishing, forestry, and trade. Improving living conditions and infrastructure in rural areas and eradicating poverty in the least developed geographical areas is critical. Due to Cameroon's abysmally low and inadequate social security and protection schemes, rural elderly are left out of development planning. They continue to rely on family and lineage ties to draw on different forms of support that are fragmenting.

8.2 Data and Methods

Based on interviews, the dilemma of older people in rural Cameroon (specifically in the North West and South West regions) is explored. The study involved in-depth interviews with 120 older men and women. The study included a diverse sample of older people across gender, ethnic/tribal affiliation, age structure, and religion. For the purposes of this study, older people are defined as those aged 60 and above. Older men and women were approached directly at various key social points—home, social clubs, churches, village development meetings, farmlands and grazing sites, local markets, and drinking places frequented by older people. The purpose of the research was explained and viewpoints sought on various questions embedded in the interview guide.

Research was underpinned by gaining knowledge of the 'worlds' of older people, and how they navigate and negotiate daily life. Patterns of care and other arrangements already in place or being fashioned for their welfare were also explored. This approach proved suitable as respondents remained relaxed and willing to 'air out' their concerns and feelings on the subject. Interviews lasted one hour on average. and some were tape recorded for those who gave consent. Interviews were guided by interview schedule with framed questions to ensure consistency and reliability. This permitted an objective appraisal on forms and levels of support. Through first-hand information and direct observations, very concrete snapshots and imagery on survival strategies were uncovered, namely an older person carrying wood, fetching water, selling wood, buying and selling basic commodities and provisions, and selling local medicines (herbal therapies). Information gleaned from respondents is presented in their own words, providing a fuller picture of individual circumstances.

Accounts reflecting support arrangements in place are pieced together and content analysed, teasing out policy implications and connections to development theory with reference to social capital. Due to variations in support mechanisms noticeable for different categories of older people, narratives from Fulani herders, a pastoralist group, are also included. Fulani who reside in remote rural settlements were interviewed alongside the President of the Mbororo Social and Cultural Development Association (MBOSCUDA)—an umbrella organisation and operational vehicle for mobilisation and pooling together of the Mbororo nomadic group. Based on field data, it is argued that different scenarios are visible in care arrangements for older people in rural communities. Fulani undertaking nomadic activities, often in remote highlands, count on lineage networks and cultural codes of conduct that guarantee care and respect for elderly. This differs from non-Fulani communities where older traditional notables rely on the largesse of the rural community through direct provisions of food, firewood, and fines obtained within village traditional councils. In presenting a heterogeneous group, this study uncovers many older Fulani whose living arrangements appear distinct, yet their survival as a minority nomadic group is complicated by frequent farmer herder confrontations over land. As this group grows numerically, they are often on a collision course with other rural communities as they search for cattle pasture, with such movements undermining care arrangements for its elderly that bank on livestock as a mechanism of social protection. The interviews were recorded through an interpreter and later transcribed to reflect views expressed. The dilemma of older people in the rural areas remains unique in relation to the range of resources they draw on; accounts reflect varying care arrangements. Dominant typology is that of older people in rural settings battering for subsistence and living off fragile nuclear and extended family ties in the provision of much needed care. This may take the form of children, grandchildren and hired helpers, drafted in to provide care. Whatever care arrangements are fashioned, they depend largely on household and other community resources garnered for their survival. Research on survival strategies of older people in the rural areas gives detailed insights into agrarian modes of subsistence. This is consonant with research elsewhere; older women in rural Ghana take up agricultural work alongside meal preparation, housework, and other community affairs (Mba 2010); others highlight how respect and reciprocity impacts on the care of older people (Van der Geest 2002), how social change underpins the welfare and livelihood of the older population (Apt 1996) with intergenerational shifts having adverse impacts on providing care (Aboderin 2006).

Given the sensitivity around the subject, some respondents were cagey and some interviews proved quite stressful. In trying to determine the amount of support provided, some questions appeared distressing particularly those relating to health profile, housing and living conditions. An older man broke into tears and pointed to his abscess reddened hand that required incision and yet he had no money to go to a hospital. On such occasions, this warranted some show of sympathy, assurance, and reassurance without deviating from the aims of the research. Such distressing accounts and other distractions may potentially affect the flow and dialogic representation between the research and respondents. As Nigel Gilbert (2001) observes, ethnographers must guard against 'going native'. Temptation of getting too close without losing one's ability for detached analysis did prove to be a huge challenge. Given the diversity of responses, it was difficult to listen in to accounts and snippets of information while maintaining an academic interest simultaneously. Being confronted with questions on government's lethargic pension policy and support for older people was tempting. The researcher had to detach himself by not passing information as it was not his responsibility and this sometimes proved frustrating for respondents.

In piecing together the accounts, the tendency to be very detailed or brief proved difficult. Respondents' accounts are presented in their own words leading to long quotes that may appear structurally deficient. However, accounts have been teased out for clarity through careful interrogation of the data. Care arrangements in rural areas are catalogued drawing on social capital theory to explain survival and coping strategies in rural settings. This in itself can be limiting as it does not capture the entire picture and degree of poverty. Case studies and narratives point to a gamut of activities that are mobilised and support networks in place which remain variable. While a few rural older people are coping, most are severely disadvantaged and at high risk of deprivation and social isolation depending on patterns of care and forms of support.

8.3 Findings

8.3.1 Care Patterns and Survival Arrangements

Older people without pensions resident in rural areas are most severely affected as access to pensions represents a direct means of income. Though pension processing and administration proves onerous, it remains vital in ensuring their sustenance. A vast majority without pensions live in abject poverty as income generated through alternative sources is inadequate. Access to pensions starkly reveals the vulnerability of older women as they constitute a small number of pensioners and indirect beneficiaries as widows drawing on a widow's allowance.

Outmigration threatens survival arrangements of older people in rural areas. Older people often struggle to cope with exigencies of rural living because family support mechanisms are fragmenting. In-depth interviews reveal the novel phenomenon of hired helpers is increasingly being fashioned as a means of providing care. The reliance on hired helpers (young boys and girls recruited by mutual agreement to provide support in return for a specified form of assistance) is increasingly used as a means of providing care. A male respondent put it starkly:

> In our days, we grew up with our children knowing that they will look after you when you are older and weak. Now, things are in reverse gear. My wife died last year and my daughter has sent a boy to stay with me. He looks after me and my daughter then pays for him to go to a government school in the village. He cleans the house in the morning, carries water, cooks my food and then he goes to school (Male, aged 68, Mbatu).

This account suggests that reliance on immediate and extended family is waning and other arrangements like hired helpers are being fashioned. These hired helpers play a crucial role in enhancing the well-being of older people. They are involved in securing essential needs such as provision of food, help with daily care (hygiene and sanitation), fetching water, assistance with farm work, accompanying frail older people struggling with ill health to various treatment centres, and offering 'shelter' from isolation, boredom, and abandonment.

In the interviews, about 65% of older men and women indicated that they preferred living in rural areas. They have a close attachment to the land and it provides them with some means of surviving. In addition, they benefit from easy access to land for farming and other agricultural activities including livestock rearing, animal husbandry, and planting of eucalyptus trees, eventually felled down for wood. This view is captured in this account:

> I have lived in this village since birth; my umbilical cord is here. I have plenty of natural attachment to this place because I survive on the land I inherited from my grandfather. I am able to rent some of the plots for farming and I can also cultivate food crops for sale and household consumption. When I am not physically strong, I can rally others to clear land, and for weeding and harvesting. Without this land, life would have been very difficult for me. I have also planted a 'raffia bush' which enables me to tap palm wine. It keeps me busy and I am able to get some days about 1500–2500 CFA. If, I had to live in town, I would have to purchase land for building and farming. Land is scarce and if you are lucky to find one, the price is too high (Male, aged 68, Awing).

A similar sentiment is expressed by another respondent:

> I prefer living in the village as I am surviving on a piece of land left behind by my late husband who was a polygamist and married to three wives. Luckily, we were allocated pieces of farmlands for agricultural activities. The land is my life-blood because it keeps the family going as we are able to farm on plots and we sell some of the produce in the local market. I also plant plantains and palms which require plenty of space. I sell plantains and palm oil extracted from the palms (Female, aged 71, Bafut).

Still another respondent considered the issue of respect and close ties that exist within rural communities:

> I think life in the village keeps us closer to nature. We still feel for one another, respect for elders is still visible and there are traditional institutions to enforce our tradition and culture. I am a village notable and the secretary of the village traditional council. We are respected in the village as we settle land disputes, boundary disputes, marital problems, traditional marriage, witchcraft and other problems within the community. When I walk around the village, villagers offer me food, drink and help with farming. I feel better here because in the town, we are losing the value of tradition and respect for older people (Male, aged 68, Mankon).

This account reveals that many older people in the rural areas receive support from the local community and survival is closely tied to participation in village and community activities. Such involvement in social events adds to the stock of social and cultural capital with tangible returns such as food and drinks and entrenching community values of respect and reciprocity. In the rural areas, the elderly articulated that they feel part of the community and some also are affiliated to particular faith-based organisation. One respondent highlights the role of faith-based organisations:

> I go to church every Sunday and this means a lot to me. Singing and listening to the scripture is healing and church members look after the old. I have received support of foodstuff, wood for heating or cooking and others visit to make sure I am okay if I do not attend church (Female, aged 73, Nkambe).

The role of faith-based organisations in providing care is pivotal. Direct provision of basic necessities such as food and pastoral visits to communities is transforming lives, making life bearable for often isolated and neglected rural dwellers.

Inhabiting rural areas with trappings of underdevelopment and a dearth of social services makes life unbearable for older people. In their accounts, they negotiate survival through a vast array of informal sector activities, farming, and other allied agricultural activities. Livelihoods are often contingent on a vast array of informal sector activities. This is captured in one respondent's account:

> I own a farm and have been growing snails and this is thanks to support from an NGO [non-governmental organisation] that helps us with skills on how to keep the snails. I sell them to traders for their business and they use it for food preparation like 'congo-meat, pepper soup and snail soya'. The money raised keeps me going and I am able to pay school fees for my children and grandchildren (Male, aged 66, Mamfe).

Yet another respondent said:

> I have been a farmer and *buyam-sellam* all my life. My deceased husband was also a farmer and we had a cocoa and oil palm plantation. When my husband died, I set up a provision store at home and I go to rural markets to buy foodstuffs which I resell in the store. In fact, I bought a diesel operated oil press machine, which I use at Mbalangi. With the oil press, I process palm oil for the store and I also charge a fee for the processing of palm oil for others. I am known at Mbalangi as Mama Diesel. My store and palm oil trade give me about 60,000 CFA a month and I am able to look after my family. I constructed a home in Fundong, I am also able to pay school fees for five children. My greatest worry is bad roads as most foodstuffs rot on the way because roads are very bad. When you are physically weak, you cannot venture into the *buyam-sellam* business (Female, aged 71, Muea).

Access to land becomes fundamental in shaping their livelihood, and without access to land, older women who often constitute the bulk of the rural elderly are hard hit and survive through individual and collective agency. It is widely accepted that transport infrastructure also has a substantial impact on economic growth and poverty. Due to bad roads, the strain of coping is evident, echoing a study by the World Bank (2009) conducted in Cameroon which emphasises road infrastructure as vital in reducing poverty.

Land is a vital resource and a strong determinant of the health and well-being of rural elderly as some accounts point to its centrality. A male respondent relates the importance of agriculture to their survival:

> We have been farmers and even now that we are old, we cannot stop working. We cultivate plantains, palm oil and other food crops. We have 150 palm trees and we extract palm oil from the collected kernel every two months. We may have a tin and half of palm oil a week. Usually, we share part of the palm oil with our children who are married and sell a tin for about 8,000 CFA. For the plantain, a bunch ranges from 500 to 1500 CFA providing 'chop money' for other things. Part of the plantain is used for feeding the household. Cassava is processed into garri which is sold by my wife in the local market every week. From the sale of food crops, we are able to contribute towards our weekly njangi [cash generating society] of 1000 CFA and monthly njangi of 1500 CFA. This is used for children's school fees, hospital costs and to buy other basic necessities like soap, rice, and other ingredient (Male, aged 78, Mankon).

Access to land and ownership remains vital assets in guaranteeing livelihoods as a respondent stated:

> In fact, I am still alive today because of the landed property I inherited from my parents. The land is rented out to farmers on an annual rate. I am able to raise 45,000 CFA every year. Added to this, I am involved in food crop cultivation and snail domestication which generates about 35,000 CFA every month. My trouble is my health as I have asthma and high blood pressure but thank God, I am familiar with local herbs to treat it (Male, aged 68, Mamfe).

The lack of running water also represents a major handicap in improving the well-being of older people. One respondent described his ordeal:

> We fetch water from a distant stream (about 10 kilometres away). Most often the water is not good for drinking as people bathe and use it for other things. During the rainy season, we mostly collect rain water and store for usage. Of course, the water itself and storage conditions make us ill (Female, aged 68, Nkambe).

A male respondent recalls his troubling experience:

> Our major headache is lack of water. We had a borehole in the village which broke down and we are always ill with typhoid which entails us going to the hospital most times and the closest health facility to us is the Baptist Hospital at Mbingo. When we have the means we have to go there and at times we rely on traditional herbs that we boil and drink (Male, aged 68, Mankon).

With the lack of basic amenities, the plight of older people resident in rural areas is summed up in the account of a widow:

> Life remains a constant struggle and even worse for us as we have to stretch and stretch to have something to eat. Things would have been better if we had good roads to easily transport our produce to local markets. We don't have piped water and this takes up much of our time as we, with the support of our children and grandchildren, trek for long distances to get water (Female, aged 68, Mamfe).

Poor water services continue to account for rural–urban disparities: though there is notable progress, in 2006 70% of the population had access to safe drinking water; coverage in urban centres is 88%, significantly better than the 47% in rural areas. Many peri-urban dwellers also lack access to safe drinking water (WHO/UNICEF 2008). The situation of older people is complicated due to erratic supplies and distance covered to get clean drinking water.

Access to health services and hospitals exacerbates the already worsening health condition of most rural dwellers. On lack of health services, a respondent declared:

> I have been suffering with high blood pressure for the last six years. I managed to go to the hospital and was given some 'aspirin'. I am still alive, thanks to my knowledge of herbs which I prepare and drink. With my knowledge of herbs, I am now a traditional herbalist offering help on herbal medicine to other villagers for a token fee. This activity and cultivation of food crops keep me going as I am able to generate some money to buy other basic necessities (Male, aged 67, Mamfe).

Another female respondent was worried:

> I am very scared if I fall sick and cannot rely on my herbs because the hospital is very far and apart from the long journey, I do not have the money to pay for the consultation (Female, aged 68, Mamfe).

This account is telling as it reveals that older people are forced to rely on herbs to treat certain ailments due to inadequate access to health facilities, which is compounded by lack of money—a major deterrent to accessing key services.

Gender underpins survival and coping strategies. Childless women and widows often constitute the more highly disadvantaged groups as they have to carry a heavy burden of fending for themselves. This is heightened by anachronistic traditions and customary laws that impede the socio-economic advancement of women. A widow recounts her survival approach:

> I am 77 years old and I have been a traditional birth attendant since the age of 21. I picked up the skill from my late mother who was naturally gifted with herbs. I have delivered about 100 children in remote villages around Mamfe. I also prepare herbs to ease the pain of labour for pregnant women and also to aid breast flow for nursing mothers. I do not charge any fees for this activity but I am highly respected and compensated by villagers with money, food crops, farming plot (Female, aged 77, Mamfe).

Women are heavily reliant on farming for their economic survival but due to anachronistic traditions they are denied access to land. A widow's account is instructive:

> I am now a rural farmer and petty trader and I grow food crops on land that I rent from others. When my husband died, his land was taken over by his relations. I sell vegetables and other foodstuffs in rural and urban markets. I usually make about 15,000 CFA every month from the sales. I am also relying on a small pension and also rent from two houses which we constructed when he was still alive. With this income, I run a household with six children (Female, aged 67, Mankon).

Overall, these quotes reveal how older people in rural areas are living on the edge of poverty as their quality of life is affected by the dearth of infrastructure and basic amenities. However, all is not bleak as older residents of rural areas enjoy tangible rewards through greater networking and heightened community relations. The quotes reveal that older people in rural areas rely on social capital to navigate life in a rural landscape.

8.3.2 Does Social Capital Count?

Social capital as contained in narratives is pooled through varied sources: participation in social, economic, spiritual, and cultural ventures. Respondents mention belonging to cash generating societies [njangis], faith-based groups, traditional associations, and kinship-based groups to provide vital support, crucial for their care arrangements. Being in a state of material poverty can be a persistent or an episodic experience that can be life threatening, life restricting, and disempowering (Dixon 2010). The accounts from the interviews show that older people are seeking to escape the grip of poverty through reliance on trusted relations within village-based organisations, church groups, family support networks, and other solidarity networks, though such mechanisms are increasingly under strain.

Though fraught with challenges, building capacity and engaging in social and traditional networks is related to social capital theory. Literature on social capital carries an idealisation of 'community' which is seen as the prime source of building social capital. A paper by Collier (1998) notes 'the quantity, quality and persistence of social interactions among neighbours, friends and members of groups and associations, generate social capital and the ability to work together for common good. This is especially important for the poor as it can be used by the poor as a substitute for human and physical capital'. Social groups organised around relations of commerce or friendship benefit from and contribute to interpersonal trust among group members. This subsequently helps strengthen the group and contributes further to its success (Bourdieu 1980; Coleman 1998).

In a general sense, social capital is rooted in repeated social interactions between individuals and groups said to develop trust, social norms, and strengthen co-operation and reciprocity (Fukuyama 1995; Coleman 1998; Warren 1999; Woolcock 2000). The amount of social capital built depends on the quality and quantity of interactions as close family ties act as the social glue guarding against vulnerability (Narayan 1997). Often overlooked in current discussions of social capital from a gender perspective is the uncritical treatment of relations within households, families, and kin groups (Mayoux 2001). It is meaningless to speak of social capital unless it is conceptualised as the sum total of social relationships and networking directed at the welfare of a group. The whole stock of social capital remains vital mechanisms of support for those trapped in poverty. For those living in deprivation in poor neighbourhoods, reliance on social capital, converted in material and financial resources, are vital 'social security mechanisms' (Von Benda-Beckman and Marks 2000). The interviews suggest that social capital obtained from the household, community, and other support networks is pivotal in efforts to stave off poverty. Njangis [social networks where cash is pooled and given to members in turn] and other family support mechanisms remain pivotal. In seeking out new ways of improving their lot, perspective of social capital lends credence but may play out differently with other categories of older people. Amongst Mbororo Fulani, kin networks and cattle transactions are key survival mechanisms.

8.3.3 Fulani Herders and Lineage Transactions

Operating on the fringes in Cameroon's rural areas are Mbororo-Fulani pastoralist groups: a growing nomadic herder community dotted across remote highland settlements. Its transhumance activity and relevance in rural monetised economies cannot be overstated as herds of cattle remain major sources of raw protein for many rural and urban populations. Hickey (2002) notes that ethnicity, religion, and to a lesser degree, race, provide the key aspects of identity that define the Mbororo as a distinct group. Fulani practise the Islamic religion enforced by a strict cultural code of behaviour and ethos known as *Pulaaku*. The code provides both a moral framework and a code of conduct for the pastoral Fulani, and is also maintained by the Fulani operating in urban areas. This code of conduct is intimately bound up with nomadic pastoralism and with good animal husbandry. It is also bound up with the fulfilment of duties to elders, wives, and the lineage group, and the proper arrangement of marriages. The four dominant strands of *Pulaaku* are identified as: fortitude in adversity and on ability to accept misfortune, sound common sense and manners, reserve and modesty in personal relations, and dignity. For the Mbororo Fulani themselves, *pulaaku* makes them unique and different. It is about dignity and hiding problems. Mbororo Fulani equates their distinctive pastoral way of life with their ethnic origin, to the extent that there is strong attachment to the lineage. The coordinator stated that there is a continued tendency to marry within migratory groups, often with close cousins, as a means of preserving *pulaaku* and herds. The ethnic exclusiveness of the Mbororo culture is a factor that often sparks conflict with their non-Fulani farming neighbours.

Pastoralism in Mbororo families is managed at the level of the household, with cattle ownership concentrated in the male family head. Livelihood roles within compounds are allocated according to gender and age. Older Fulani are responsible for all aspects of decision-making and activities regarding the movement, health, and sale of cattle. The wife or wives of the family head have milking rights, but lack the power to sell the cattle. An elder or older person amongst the Mbororo is referred to as *Nyako* and they command a great deal of respect as prescribed by *Pulaaku*. Younger Fulani owe the *Nyako* respect and are expected to assist with basic needs like fetching water, washing of clothes, dispatching messages, and looking after herds of cattle. On the same scale as *Nyako* is the *Ndotijo* family head, this status is conferred on persons within the nomadic community who go through daily challenges like raising children and managing large herds of cattle. The project coordinator of MBOSCUDA revealed that his organisation is battling to work out support for frail Fulani. They are feeling the pinch of youth migration as young Fulani now strive for western education to improve their lot. The coordinator explained that attempts are made to factor in projects that may indirectly benefit the elderly:

> If the older people have children in school, they benefit from the sponsorship programme for children especially if such older persons are not in a good financial standing. Also, if they reside in communities where some of our projects are undertaken, their wives can benefit from literacy programmes and micro credit schemes, thereby relieving them of some of these burdens.

Though fraught with economic and other developmental problems that may project these communities as backward, a Fulani older male respondent sums up their care arrangements and plight.

> We are constantly on the move because of the threat of being chased away from land. We cannot enjoy a good sleep because we think about our families and our cattle, which we depend on as our only means of survival. How can our children even assist us when other groups look down on us? When we talk of care for older people, we think about our cattle first because we and our families eat from there (Male Fulani herder, aged 56, Sabga).

Another herder explained:

> As older people, our cattle is the only form of support that we pass on to our children but our neighbouring farming communities are making things difficult for us as there are always problems. We are accused of cattle trespassing and destruction of food crops. Our children are frightened and running away from us (Male Fulani herder, aged 69, Nkambe).

One herder also stated:

> We, Mbororo do not enjoy any support from the state. Our children do not have public service jobs, we make a big contribution to the economy through our cattle activities, yet we do not have any social amenities like water, health centres and electricity within Fulani settlements even though our cattle are heavily taxed. We are always duped by administrative authorities when there are farmer grazer problems as we are looked at as a primitive minority people without access to land. We offer ransoms to these authorities and yet the problem resurfaces after a short while without any solutions. Even to get access to the judiciary to present our case is not possible as we are always considered as intruders. Cattle rearing remains our soul and blood, everything about us: history, identity, social security, culture and mode of subsistence and still, we do not have any protection from the state' (Male Fulani herder, aged 71, Sabga).

8.4 Policy Implications

In uncovering the social reality and care arrangements of older people in rural areas, closer scrutiny is needed of the context of poverty and economic survival strategies. A starting point is a more detailed survey of the demographic, economic, and social conditions of older rural dwellers. Research to inform policy, effective planning, and targeted resource allocation for an ageing rural population cannot be delayed. Periodic research to unravel the impact of population ageing in rural areas cannot be overstated given the ramifications on policy formulation and development planning.

As case studies and scenarios show, one size cannot fit all in terms of policy articulation given variations and categorisations in levels of support drawn upon by rural dwellers. What works for a female farmer aged 65 years and over may impact adversely on a male of the same age; the same policy framework may not match the needs of Mbororo Fulani communities who battle with cattle and other transhumance activities to scrape a living. Arguably, all categories need a critical range of basic amenities such as roads, piped water, electricity, and rural hospitals to make life bearable. More survey data is needed to highlight the plight, contribution, and challenges faced by the rural older population.

Rebuilding rural infrastructure is crucial in livelihoods improvement as most older people mention lack of basic amenities such as roads, health care, good drinking water, and electricity impacting negatively on coping and care arrangements. A practical approach would be greater government collaboration in supporting village development associations and faith-based organisations. These are development oriented, yet lack the needed capital and injection of vital funds. As the ageing of the population has ripple effects on other sectors, it is imperative for the Ministry of Social Affairs to better streamline and engage in fruitful partnerships with other government departments in fashioning a policy to safeguard the population in rural areas through targeted development schemes to rebuild rural infrastructure to better harness the potential of older people. A starting point is to identify and work with non-governmental organisations addressing concerns of older people. Rural community festivals and cultural events celebrating the virtues of older age would revitalise social networking and social capital and offer recognition to older people in rural areas as most remain potent forces in economic development.

A vast majority of the rural older population depends largely on agriculture for their survival. The need for these activities to be revitalised for economic development can no longer be overlooked. A take off point is provision of financial incentives, small credit, farming implements and updates on agricultural innovation and farming techniques. Further, organisation of agricultural shows for older farmers could foster exchanges on best agricultural practices and yield greater dividends in terms of opportunities for active ageing. Working laterally with non-governmental organisations like Community Development Volunteers for Technical Assistance (CDVTA) and others very active in the regions under study is most welcome.

The rural older population should be more proactive in forming rural clubs to promote social capital ventures (like njangis) and farming clubs (like gardening, bee farming, tree planting, herbal therapies) to improve marketing and networking. Such clubs can be used by health, agricultural, and social welfare officers to transmit best practices and new information on nutrition, health screening, and tips as well as targeted social assistance. Such clubs would build psychosocial support, ward off isolation, and promote mental stimulation.

A gender consideration in ageing policy is vital due to the differential impacts on care arrangements and livelihoods. Research and policy capturing differences and disparities is urgent, as are empowerment mechanisms for most vulnerable groups like widows, widowers, childless women, chronically ill, disabled and frail rural elderly. Training of social workers and social work assistants to address the needs of the most vulnerable trapped in poverty cannot be delayed.

Embedding perspectives of affected groups in care planning would enable voices of rural people to be heard in terms of their priority needs in development planning. The Ministry of Social Welfare working in collaboration with other Ministerial Departments should conduct periodic assessments and reviews to determine changes in care arrangements and policies to tackle imbalances. Focusing on nutrition, hygiene, and sanitation programmes, water treatment and purification strategies hold the potential to transform living standards of the rural older population.

8.5 Conclusion

From a snapshot of social realities of older people living in rural areas, this qualitative study has articulated variations in levels of care and support. With changing demographics and ever-shrinking national budgets for social programmes and lesser government support, older rural dwellers are left to their own devices. They suffer disproportionately due to less disposable income and are battling to survive on traditional support systems that are in a state of flux. The interviews paint a picture of greater bonding, social cohesion, respect, and strong social networks as vital assets. While outmigration remains a strong determinant of care arrangements, this study has highlighted the different scenarios and support mechanisms that are being fashioned to provide care for older people with individual, yet unique experiences. The findings point to family resources being over-stretched and hired helpers becoming key players in the rural care economy. Life chances are further undermined by a dearth of rural infrastructure and lack of basic social and health amenities that are impacting care arrangements being fashioned. As portrayed in narratives from the interviews, rural elderly are increasingly banking on social capital with tangible (*Njangi*) and intangible benefits (solidarity groupings, close knit support networks, respect), support from faith-based organisations, and the voluntary sector as additional sources of support. While Fulani nomadic communities look to kin and lineage arrangements regulated through *Pulaaku*, this is under pressure due to limited input by younger Fulani in cattle transactions. The search for identity and integration continually puts this group on a collision course with other rural communities, thereby jeopardising efforts to provide support for older Fulani. As population ageing escalates, providing care for older rural dwellers becomes taxing. Improving living conditions require a medley of infrastructural development projects (roads, health care and sanitation, access to water), direct social assistance schemes (food and essential goods), voluntary sector contribution, and streamlined cooperation with faith-based organisations and community development associations.

References

Aboderin, I. (2006). *Intergenerational support and older age in Africa*. New Brunswick, NJ: Transaction Publisher.

Abwa, D., & Fonchingong, C. (2004). Gender and poverty alleviation in Cameroon: Social equation and cause for concern. In J. Endeley, S. Ardener, & R. Goodridge (Eds.), *New gender studies from Cameroon and the Caribbean*. Oxford, England: African books collective.

Apt, N. (1996). *Coping with older age in a changing Africa: Social change and the elderly Ghanaian*. Aldershot, England: Ashgate.

Bailey, C., & Turner, C. (2002). Social security in Africa: A brief review. *Journal of Aging and Social Policy, 14*(1), 105–114.

Bourdieu, P. (1980). Le Capital Sociale: Notes Provisoires. *Actes Recherches: Science Social, 31*(2), 2–3.

Coleman, J. (1998). Social capital in the creation of human capital. *American Journal of Sociology, 94*(Supplement), S95–S120.

Collier, P. (1998). The political economy of ethnicity. Paper presented at the Annual World Bank Conference on Development Economics. Washington, DC, April 20–21.

Darkwa, O., & Mazibuko, F. (2002). Population aging and its impact on elderly welfare in Africa. *International Journal of Aging and Human Development, 54*(2), 107–123.

Dixon, J. (2010). Poverty: An existential-humanist perspective. *Poverty and Public Policy, 2*(4), 111–129.

Fonchingong, C. (2005). The mechanics of communitarianism and social capital in North West Cameroon. *International Development Planning Review, 27*(4), 427–449.

Fukuyama, F. (Ed.). (1995). *Trust: The social virtues and the creation of prosperity*. New York: The Free Press.

Gilbert, N. (2001). *Researching social life*. United Kingdom: Sage Publications.

Hickey, S. (2002). Transnational NGDOS and participatory forms of rights-based development: Converging with local politics of citizenship in Cameroon. *Journal of International Development, 14*, 841–857.

ILO. (2001). *Social security: A new consensus*. Geneva: ILO.

International Fund for Agricultural Development (IFAD). (2008). Rural poverty in Cameroon: Resource document. http://www.ruralpovertyportal.org/web/guest/country/home/tags/cameroon. Accessed Oct 2011.

Kalasa, B. (2001). Population and aging in Africa: A policy dilemma? Paper presented at the International Union for the Scientific Study of Population's XXIV General Population Conference. Salvador de Bahia, Brazil, August 18–24.

Mayoux, L. (2001). Tackling the downside: Social capital, women's empowerment and microfinance in Cameroon. *Development and Change, 32*(3), 435–464.

Mba, C. J. (2010). Population aging in Ghana: Research gaps and the way forward. *Journal of Aging Research, 2010*, 672157.

Ministry of Economy and Finance, Direction de la Statistique et de la Comptabilité Nationale. (2002). Conditions de vie des Populations et de la pauvrete au Cameroun en 2001. Rapport principal de L'ECAM II. MINEFI, Cameroun.

Narayan, D. (1997). *Voices of the poor: Poverty and social capital in Tanzania*. Washington, DC: World Bank.

United Nations. (2007a). World population prospects: The 2006 revision. http://www.un.org/esa/population/publications/wpp2006/FS_aging.pdf. Accessed 13 Mar 2011.

United Nations. (2007b). *World economic and social survey: Development in an aging world*. New York: United Nations.

Van der Geest, S. (2002). Respect and reciprocity: Care of elderly people in rural Ghana. *Journal of Cross-Cultural Gerontology, 17*, 3–31.

Van Ginneken, W. (2003). Extending social security: policies for developing countries. *International Labour Review, 142*(3), 277–294.

Von Benda-Beckman, F., & Marks, H. (Eds.). (2000). *Coping with insecurity: An 'underall' perspective on social security in the Third world*. Yogyakarta: Pustaka Pelajar and Focaal Foundation.

Warren, M. (1999). Democratic theory and trust. In M. Warren (Ed.), *Democracy and trust*. Cambridge: Cambridge University Press.

WHO/UNICEF. (2008). Coverage estimates: improved sanitation, Cameroon. Joint monitoring programme for water supply and sanitation. http//documents.wssinfo.org. Accessed 12 May 2011.

Woolcock, M. (2000). *Friends in high places? An overview of social capital* (ID21 Insights Issue #34). London, UK: DIFD.

World Bank. (2009). *World development report 2009: Reshaping economic geography*. Washington DC: World Bank.

Chapter 9
Ageing in Post-Apartheid South Africa: An Analysis of the "Nearly Old"

Nina Hunter and Julian May

9.1 Introduction

Poverty is a concern among the older population in many countries, but especially in sub-Saharan Africa in which both the proportion and the numbers of poor have continued to rise throughout the past 3 decades (Chen and Ravallion 2008). A key factor of older people's poverty is their reduced capacity to work, resulting in lacking the income to either sustain a healthy lifestyle or access health care. This set of circumstances leaves them vulnerable to many health conditions associated with ageing as well as to the shocks normally associated with poverty.

Although a relatively new concern in the literature relating to poverty in developing countries, the plight of older people has been documented in some regions (see Ferreira 2000; Ferreira et al. 1992; HelpAge International 1996). However, research has tended to focus on those aged 60 and over, reflecting the retirement age applied in many developed countries and the age at which benefits such as state pensions or subsidies become available. This eligibility determined cut-off does not necessarily represent when ageing actually becomes a source of vulnerability. In fact, in Africa the more traditional definition of older persons correlates with the chronological ages of 50–65 years, depending on the setting, the region, and the country (WHO 2011).

For this reason, this chapter shifts attention to a group of individuals who have not received explicit attention in the literature on ageing: what will be termed the "near-old", aged 50–59, who are preparing for the transition to old age. This can be thought of—at least for some—as a period in which there is reduced productive activity, whether voluntarily or not, and in which reproductive activities have been or are being finalised. However, it is also potentially a period of high vulnerability in which negative events or shocks such as retrenchment, injury, or ill health can

N. Hunter (✉) • J. May
School of Built Environment and Development Studies,
University of KwaZulu-Natal, Durban, South Africa
e-mail: ninahunter@telkomsa.net; jmay@uwc.ac.za

P. Maharaj (ed.), *Aging and Health in Africa*, International Perspectives on Aging 4,
DOI 10.1007/978-1-4419-8357-2_9, © Springer Science+Business Media New York 2013

undermine prospects for a successful transition from working life to retirement. From the perspective of public finance, individuals in this age group are the next generation of older people, and future resource requirements can be estimated from the numbers and situation of those in this age group. Failure by this group to accumulate sufficient resources to sustain their retirement from economic activities will increase their demand for support from public resources. If this is not forthcoming, they will be compelled to liquidate assets that they may have accumulated and eventually turn to support from their children, thereby increasing the likelihood of the inter-generational transmission of poverty.

South Africa provides a useful case study for this examination. By definition a middle-income transition economy, it has the highest percentage of older persons in Africa (Lombard and Kruger 2009) and one of the most rapidly ageing populations in Africa. In 2001, 7.3% of the population was 60 years of age and above, and 0.9% was 80 years and above (own calculations using data from Statistics South Africa 2003). In the year 2050, 15.3% of the population is projected to be over the age of 60, while 2.3% of the population will be over the age of 80 (own calculations using data from the United Nations Population Division 2008).

The majority of South Africa's older population is African, female, income poor, and lives in large, multi-generational, female-dominated, and female-headed households that tend to be concentrated in rural areas. Most older people stay in households with younger relatives (multi-generational households) and many live in skip-generation households (Lombard and Kruger 2009; Statistics South Africa 2010b). Households headed by older persons that include children have the highest total dependency ratios of all categories compared. Older persons have to support almost 1.5 children for every older person in older female headed households (Statistics South Africa 2010b).

What distinguishes the older population in South Africa from many in sub-Saharan Africa is the state Old Age Pension (OAP)—a non-contributory grant funded from general government revenues—which, as of 1 April 2010 both females and males receive as entitlement at age 60. African older people are "the apartheid-era generation who were not educated nor incorporated into formal labour to provide for their retirement" (Lombard and Kruger 2009, p. 123). As such, they continue to suffer the legacy of discrimination and an inadequate education (Carter and May 2001; Statistics South Africa 2010b). The OAP therefore provides an essential means of support in old age, with 70% of older persons receiving some kind of government transfer in 2009 compared to only 28% of the general population (Statistics South Africa 2010b, p. 99). The pension is an important means of support not only for the pension recipient but for the households in which they live (Ardington and Lund 1995; Lund 1993). More recent work (Burns et al. 2005) confirms the key poverty protection role of the pension among households with older persons in which younger unemployed members are supported by the OAP. As an illustration, Barrientos (2003, p. 17) shows that the poverty headcount would be 2.8% higher in South Africa if the pension income was removed, and the poverty gap would be 40% larger. The OAP also facilitates household access to basic services and economic opportunities, but overall, the OAP targets household poverty rather than facilitating older persons' independent living (Lombard and Kruger 2009). Indeed Eckley (1996, as cited in Lombard and Kruger 2009, p. 126) describes the demand on African social pensioners to share their monthly pension with unemployed family and children as a feature unique to South African society. For these reasons, those receiving an OAP have been classified not as consumers but as "societal investments" (Lombard and Kruger 2009).

Receipt of the OAP does not mean that older people retire from economic activity; indeed many are forced to seek additional income. Older people are reported to remain in the labour force in a variety of ways, namely business ownership, casual work, paid domestic work, unpaid family work and farm work, amongst others. Moreover, as caregivers, older people can in certain circumstances access additional grants, such as the care dependency grant, the disability grant, and the child support grant (CSG) (Lombard and Kruger 2009).

If anything, most older people in South Africa are still performing (re)productive activities, something that the AIDS pandemic is largely responsible for. The pandemic has contributed towards the erosion of the traditional extended family, it has taken children or other relatives who would have cared for older people, and it has impacted on families' ability to care for older people. In addition, services in South Africa have shifted from primarily institutional care to community-based care. Since so many of South Africa's older people are burdened in their role as caregivers, they could be said to be experiencing "a retirement lost" (Lombard and Kruger 2009, p. 132). In addition, their health status is also at risk as a result of AIDS, because of their caregiving activities as well as their own sexual behaviour (Lombard and Kruger 2009, p. 132). The prevalence of acute and chronic diseases increases with age, and older people are more likely to experience compromised health than those at younger ages, although they also tend to be less protected against unforeseen medical expenses (Statistics South Africa 2010b).

The near-old (50–59 years) are a substantial sub-group of older people in South Africa. According to data from the 2001 census, 6.3% of the population were near-old (Statistics South Africa 2003, p. 27) and 11% of the population will be 50–59 years of age in the year 2050 (United Nations Population Division 2008). Near-old age thus represents an important issue for investigation in which opportunities might still exist to mitigate poverty among older people. Moreover, since South Africa is one of a small number of developing countries to have a well-functioning state pension system, it is an interesting case study for any investigation of the situation of the near-old as receipt of the OAP represents a significant change in their economic status.

We use a life course framework to investigate the employment, income, and health situation of the near-old. Earlier research (see Hunter and May 2003) employed large-scale household surveys undertaken in the late 1990s to calculate descriptive statistics which highlighted the lack of differences between the near-old age group and those at older ages. This analysis suggested that near-old age was an important issue for investigation, since opportunities could still exist to prevent poverty among those at even older ages. This chapter uses data from the first wave of the South African National Income Dynamics Study (NIDS) conducted in 2008 to compare the employment, income, and health situation of the near-old with that of those at younger and older ages. We explore whether this group is likely to be more vulnerable than the traditionally understood older people because they face unemployment at an age when their chances of finding employment are already diminished, due at least in part to diminishing health status, but are not yet eligible to receive a state pension. However, in any such analysis in South Africa, the situation of Africans has to be separated from that of Indians, coloureds and whites. This is because apartheid and the legislation and institutions through which this ideology was implemented operated to produce persistent poverty and extreme inequality (Hunter et al. 2003).

To begin we review the literature on poverty and old age, as well as the unemployment of older people. In addition the interrelationship between health and poverty, particularly in old age, is also explored. Next follows a discussion of some of the themes of vulnerability as old age is approached. We then explain how older people are defined in this chapter using a life course framework to inform this discussion. While the use of chronological age is not without problems, some means of defining older people is necessary for the quantitative data analysis that appears later in the chapter. The remainder of the chapter is taken up by an analysis of the data from the 2008 NIDS which elucidates issues of poverty, health, and well-being with a particular focus on the near-old and how they compare to others in older and younger age categories. Finally we draw some conclusions, suggest avenues for further research and possible policy approaches.

9.2 Poverty, Unemployment, and Health of Older People

In the context of an "ageing world", unsurprisingly, older workers have moved up the policy agenda within the industrialised nations (Kinsella and Velkoff 2001). As an example, although in the 1980s and first half of the 1990s, policy-making in much of the European Union emphasised the virtues of early retirement, since the late 1990s, there has been an increasing emphasis on overcoming age barriers in the labour market, closing off "early exit" pathways and promoting the employment of older workers. This has been driven by concerns over ageing, shrinking labour forces, and the sustainability of public pension systems (Platman and Taylor 2004; Taylor 2004).

However, although the participation rate of older workers, rising since the end of the 1980s, is projected to continue increasing, the participation rate of older workers' is still only half the participation rate of the prime-age group (25–54 years of age) (Chen et al. 2009), and unemployment of older people continues to receive attention in developed country literature. Various authors (including Price et al. 2004; Encel and Studencki 2004) identify age as the single most important barrier to obtaining employment for older workers. Defined as "differential and discriminatory treatment on the grounds of age" (Price et al. 2004, p. 182), age discrimination in the labour market has been identified as a primary factor that acts to limit not only employment opportunities available to older workers, but also training opportunities and conditions of employment. Many mature aged job seekers report being forced into early retirement through retrenchment and/or redundancy, or being pressurised into accepting part-time, casual or contract work that holds no sustainable or foreseeable long-term employment prospects (Price et al. 2004). For others, part-time work may be used as a gradual transition to retirement (Walker 1999).

Health deterioration can also lead to a lessened ability to work, and this is combined with the tendency for real wages to decline later in the life cycle (Raymo and Cornman 1997). Apart from this, the employment of older people is faced with a number of challenges. Employing older workers is considered by some to be a barrier to employment—both perceived and real—since employers would rather hire a younger person at a lower salary with lower health care costs. Powell-Griner et al. (1999) describe how companies providing health insurance coverage are less likely to hire older workers, who are more costly to insure. In addition many older workers seem to desire to phase down work, while some older workers may need ergonomic adjustments to accommodate their changing abilities and physical conditions (Chen et al. 2009). Further barriers include seniority pay scales which make it more costly to hire or retain older workers (Jackson and Howe 2009) and over-rigidity in employers' approaches (Loretto and White 2006).

Experience is mixed with regard to developing countries. Barrientos et al. (2003) report that the rates of participation in the labour market by older persons are higher in developing countries largely due to the absence of alternative forms of support. However, a study of older persons (aged 60 and over) in Vietnam (Friedman et al. 2001) found that this group was less likely to be working to the extent that they are older or sick, although there is little variation by socio-economic status. There were, however, distinct patterns of work stoppage across employment sectors (agricultural, state, and non-state), due in part to the nature of work in these sectors. The data demonstrates the strong outflow of older employees from the state sector.

These differences reflect the diverse contexts of work and retirement. Pensions are rare outside Europe and North America, and in most other parts of the world work remains essential for survival for the majority (Wilson 2000). While loss of wage employment is likely to lead to poverty for many older people, for some where wage employment stops, other types of employment begin. Facchini (2009) notes that in the context of widespread "informal" economy, for those who have competences which can be used in these areas, the end of "formal" working activity can bring with it the passage to a more blurred working situation and the addition of income gained from "informal" work.

For some of those no longer in wage employment, pension or informal work can provide needed income and unemployment need not result in poverty. However, while poverty has been substantially reduced amongst individuals ages 65 and older in most rich countries over the past 40 years, pensioner poverty has not been eradicated, particularly in the English speaking nations (Smeeding and Sandström 2005). The experience of this poverty among older people is gendered: older women are twice as likely as older men to be living in poverty, according to Vartanian and McNamara (2002). These authors used the 1968–1997 United States Panel Study of

Income Dynamics to examine the poverty situation of older women. The findings suggest that characteristics of women aged 40–59, such as workforce participation, income, and rural residence are strongly related to economic outcomes in later years (at ages 66 plus). Events and characteristics of women at these older ages, however, contribute significantly to economic outcomes even when midlife factors are incorporated into the analysis. In addition, both midlife and later life characteristics contribute to the persistence of poverty in old age, with many groups of poor women finding themselves in poverty for over 50% of their old age years (Vartanian and McNamara 2002, p. 532).

According to Choudhury and Leonesio (1997), marital status changes are particularly important to unravelling the causes of poverty among aged women. In fact, the greater risk of poverty for older women is closely associated with marital status (Shaw and Lee 2005). Vartanian and McNamara (2002) found that across all the periods measured in their study, married women have consistently lower poverty rates than either widowed women or the divorced, separated, or never married group. Moreover, because of longer life expectancy for women and earlier age of marriage, women are more likely than men to experience the loss of their spouses and live alone in old age (Shaw and Lee 2005). The experience of widowhood is associated with a decline in economic well-being for both younger and older women. Vartanian and McNamara (2002) cite research that shows that widows are more likely to experience a spell of poverty in old age than intact couples, and the death of a spouse is strongly associated with an increased risk of poverty for older women. Being widowed and living alone is also significantly associated with more time in poverty. Likewise, Hungerford (2001), in a study of the economic well-being of widows in Germany and the United States, finds that most widows in both countries experienced a decline in living standards, and many actually fell into poverty at widowhood, with a fall in social security and pension income being the largest contributor to the fall in living standards. Vartanian and McNamara also find that although widows have been the focus of many previous studies, divorced, never married, and separated women were as likely or more likely to experience negative economic outcomes in old age as widows. In addition, both family income-to-needs and the percentage of time in poverty in old age is affected for such women, both younger and older.

Choudhury and Leonesio (1997) also show the strong and statistically significant role of earlier life economic well-being on the poverty status of older women. However they attribute less of a role to late life events than the previous authors. They argue that most women who experience traumatic events in their later years do not necessarily become poor, and that rather financial resources available in old age, in turn, depend very much on economic status, prior to old age. A further point to remember is the fact that poverty in old age tends to be most heavily concentrated among already vulnerable groups of women such as the previously poor and the unmarried (Vartanian and McNamara 2002).

Much has also been written about the association of health status and socio-economic status. According to Ardington and Case (2009), health status and socio-economic status are important determinants of individuals' well-being.

Information on income alone, or on health alone, is not sufficient. Better health can lead to higher income, and higher income can lead to better health, but neither process can be understood without the other.

Alavinia and Burdorf (2008) undertook a study to determine the associations between different measures of health and labour market position across ten European countries. They studied 11,462 participants of the Survey on Health and Ageing in Europe who were 50–64 years old. Among employed workers 18% reported poor health, whereas this proportion was 37% in retirees, 39% in unemployed persons, and 35% in homemakers. Perceived poor health was strongly associated with non-participation in the labour force in most European countries. In addition, a lower education, being single, physical inactivity, and a high body mass index were associated with withdrawal from the labour force (Alavinia and Burdorf 2008).

In an Australian study it was found that about 50% of men and some 20% of women retire early as a result of ill health. There is lower labour force participation among persons with poorer health and these people move out of the labour force at a faster rate as they age. A range of measures suggests some decline in health in the pre-retirement age group (those aged from 40 to 64 years). Better health may therefore be a facilitator of greater labour force participation (Schofield et al. 2007).

Finally, Powell-Griner et al. (1999) analysed data collected in the United States for the period 1993–1996. They showed that 35% of the population aged 55–64 was not in the workforce due to retirement, illness, or disability. Only 13% worked part time, and 9% worked full time for part of the year. In addition to low employment rates, around 10% of this age group lacked health insurance, despite the fact that they were at greater risk of poor health (Powell-Griner et al. 1999, p. 884). Moreover, those aged 55–64 are more likely than younger adults to be in fair or poor health, to be separating from the workforce, to face high individual health insurance premiums when coverage is available, and to risk financial hardship if they incur a major medical expense (Powell-Griner et al. 1999). In short, they are more vulnerable than those at younger ages both in terms of health and in terms of socio-economic status.

There are several pathways through which vulnerability may be experienced in later stages of the life course, with access to wage employment perhaps being the most important of these. Aliber (2001) mentions the problem of lowered chances of finding employment, an issue that seems pertinent in the South African context with its much cited unemployment problem (Kingdon and Knight 2004, 2007; Leibbrandt and Woolard 2001).

The likelihood of many of the poor near-old working would seem low: for July to September 2010, only 14.4% of the labour force aged 45–54 and 6.4% of those aged 55–64 were employed, compared to 25.1% of those aged 25–34 and 21.3% of 35–44-year-olds (Statistics South Africa 2010a, p. 31). Unemployment is even more of a concern for women than men (32.2% for men but 42.9% for women using the broad definition) (own calculations using data from Statistics South Africa 2010a) since women live longer on average and require greater support for their old age than men.

Carter and May (1999) uncover three dimensions of the poverty problem in South Africa that are applicable to the near-old and which potentially constitute poverty traps that become increasingly likely over the life course. Firstly, returns to uneducated labour are very low, meaning that claims on other economic or social assets are necessary to move towards and beyond the poverty line. If these assets depreciate with time, then incomes are likely to fall. Secondly, poverty is not only a matter of few assets, but also of constraints to effective use of those assets, with the lack of access to financial assets being a specific constraint. This is especially important for people preparing for retirement since the limitation is both on access to savings facilities as well as credit. Thirdly, the burden of meeting basic needs, especially water and fuelwood collection in rural South Africa creates a "time poverty" that further constrains households' ability to effectively employ those resources to which they do have access to in the generation of livelihoods. If accessing basic needs depends upon physical strength or the availability of family labour with this strength, then incomes will fall or needs will go unmet.

Carter and May (2001) discuss this inter-relationship between time, asset accumulation, and exposure to risk. Noting that time brings both the opportunity to build and use assets, as well as negative events such as illness or job loss, Carter and May (2001) show that a segment of the poor may find themselves trapped in poverty. For those near-old who have been unable to accumulate sufficient assets to insure themselves against negative events, time may simply not be available to recover from shocks, and a temporary reversal might become permanent. As already mentioned, the time remaining to them is instead likely to carry greater risk of negative events such as ill health and loss of employment. This results in a persistence of poverty which may transfer across generations. "Near-old age" thus represents an important issue for investigation in which opportunities might still exist to mitigate poverty among older people.

Another issue is the rising costs of living over recent decades and especially in the period since 2008 in which the price of food has seen a dramatic increase. This has direct implications for those approaching old age. Mehta and Cheung (1997) and Mehta (1999) document the concerns of the near-old in Singapore over rising costs of living, specifically medical costs. In terms of provision for retirement, Mehta and Cheung point out that the cost of living had risen rapidly over the decade which preceded the study, and therefore what was seen as a large amount of savings then would fetch far less today. With escalating medical costs, any private savings that had been made could easily be "wiped out" in one catastrophic health expenditure that the household could incur (Xu et al. 2003).

A final salient issue for the near-old in sub-Saharan Africa is the caring and providing role undertaken for those who are orphaned and ill with HIV/AIDS. South Africa has the largest population living with HIV in the world, which makes this issue highly relevant to the near-old. South Africa's HIV prevalence rate for 2010 is estimated to be 19% using the ASSA 2003 model (Day and Gray 2010, p. 257). Moreover, in 2010, 4% of the population was comprised of AIDS orphans (Statistics South Africa 2010c). Increasing numbers of older caregivers are under serious financial, physical, and emotional stress due to their caregiving responsibilities.

In this context, attempts to deal with the illnesses that result from HIV/AIDS will add to the medical emergencies noted by Mehta and Cheung (1997). Although such reproductive activities would usually decline as pensionable age is approached, in countries in which there is a high HIV prevalence, this will have to now increase in order to support dependants. If available income is not sufficient, savings may have to be used or assets sold (Duflo 2000).

9.3 Analytical Framework

If the near-old have been compromised in prior life stages—either in health, or economically—it can be expected that they will be less prepared for any challenges that they face in the future. A life course view is therefore helpful in an analysis of the near-old. A life course is defined as, "a sequence of socially defined events and roles that the individual enacts over time" (Giele and Elder 1998, p. 22). Life course theory suggests the importance of the interaction between demographic, socio-structural, and cultural factors in shaping age patterns and norms. Hence, a central focus of this approach is social time and the changing social roles and events along the social trajectory which allow for differentiation between life stages (Elder 1975). Emphasis is also placed on the importance of periods of transition between life stages and the links between the stages (Crosnoe and Elder 2002). Moreover, the approach recognises the importance of generational rather than age group differences and the inter- and intra-generational transfers which occur (Elder 1975). The transition from the economically active age groups to old age is just one part of the life course.

Various components of the life course include family life (childhood development, sexual debut, timing of marriage, first and last birth, orphanhood, death of spouse, establishment and disintegration of the family unit), learning life (enrolment, progression of education, drop-out, completion of school, entry into higher education, life-long learning), and economic life (entry into the labour market, job search, job changes, earning cycles, temporary withdrawal, preparation for retirement, and retirement). Accumulation paths, social trajectories, and time-based livelihood and coping strategies are other factors to take into consideration, as well as the sequencing and transitions between stages. There are a number of advantages of adopting a life course theory, but perhaps the most relevant for this chapter is the fact that poverty is often associated with key life course transitions. In addition, this approach is useful when examining the situation of the near-old because demographic, social, and historical time places limits on opportunities and trajectories.

Indeed, Heslop and Gorman (2002) note that for the majority of the world's older people, the meaning of old age is not a chronological definition but rather the changing roles that accompany physical change and reduced capacity to contribute or maintain a livelihood. Being old can occur at different chronological ages, determined by the prevalent socio-cultural milieu, or even by the specific context of subgroups within society. Economic structures also play a role in terms of the type of work that can be carried out, the availability of health facilities and other support infrastructure, and the existence of insurance and financial markets that permit savings for the future. While some developmental processes are time dependent (such as biological growth), and as such will respond fairly well to the index of age, others may be time independent (such as physical illness) (Kimmel 1974). In line with Heslop and Gorman, WHO suggests that in Africa it seems to be more appropriate to use a combination of chronological, function, and social definitions of old age (WHO 2011). While the life course framework appears to be the ideal approach to apply theoretically, it presents analytical constraints since chronological age is argued to not always be a meaningful indicator of development or the ageing process. Elder (1975), who played an important role in the development of the life course framework, suggests that chronological age be used as an estimated guide to stage of life in the ageing process, and for the purposes of data analysis in this chapter, different "categories" of people have been defined. Although, recognising the limits of chronological age, we have chosen to follow Elder (1975) for the purposes of this chapter and propose the following categories.

9.3.1 The Young Years (15–29)

Those in this age band are likely to have undertaken sexual debut, and, for many females, first pregnancy, the completion of secondary schooling, and possibly formal education. According to O'Higgins (1997, p. 63)

> The operational definition of youth, or young people, varies widely from country to country depending on cultural, institutional and political factors. In industrialised countries, the lower age limit usually corresponds to the statutory minimum school-leaving age; the upper limit tends to vary more widely. Within the category of young people, it is also important to make a further distinction between teenagers and young adults, since the problems faced by these two groups are quite distinct.

South African labour market definitions class the youth as age 15–35. The United Nations define the youth as between the ages of 15 and 24, while the Commonwealth definition is age 15–29. Taking all of this into consideration, those aged 15–29 have been defined as the youth for the purposes of this study.

9.3.2 The Middle Years (30–49)

All in this group would have completed their own formal schooling, and most will have entered the labour market, even if this means self-employment or subsistence agriculture. In addition, most will have accumulated some economic or physical assets whether housing stock, durables, or livestock. Most will be in, or will have been in a union, and most spouses or partners will be alive. Most will have more than one dependent, some of which will be schooling and as such most will be caregivers themselves. In South Africa many will be caregivers in receipt of CSG payments, but for some the CSG will be lost as children age out of the system. Indeed, some of this group might have grown up dependents, and be losing household members. Most will have living parents, and thus may have living older dependents, or in South Africa, be drawing on the incomes of older people as a household livelihood strategy. Finally, an older household member will be designated as the head of household for a significant portion of this group.

9.3.3 Near-old Age (50–59)

Those in this age group are preparing for the transition to old age and experience many of the life-cycle changes that will lead to the success of this transition. This can be thought of as the beginning of a period of reduced productive activity, the completion of reproductive activities, including the care and education of the next generation, and the consumption of accumulated savings. The near-old are those for whom the consolidation and preservation of their assets is critical if they are to avoid poverty as they age.

It is hypothesised that although many of those in this age group are likely to be breadwinners, key decision makers, and caregivers, this is a period of high vulnerability, in which shocks and the failure to accumulate can have dire long-term consequences. Widowhood will be increasing for women along with the risk of retrenchment for both women and men. Secondly, for planning purposes, those in this age group are the next generation of older people, and future resource requirements can be estimated from the numbers and situation of those in this age group. This would include projections of the demand for social security payments, facilities, and institutional care. Finally, as noted at the outset, in Africa the traditional definition of an older person correlates with these ages (WHO 2011). In most poor economies the burden of survival puts people at a greater disadvantage than in richer countries, leading them to age prematurely. Further, in most African traditions the title "older person" is socially dictated by one's role in society—an example of this is becoming a grandparent. Such titles are conferred regardless of chronological age (WHO 2002). The choice of 50 years accommodates these definitional complexities and makes place for cultural, functional, social, and chronological categories of old age.

9.3.4 Early Old Age (60–69)

In the context of South Africa, Sagner (2000) suggests that pensionable age marks a biographical orientation point by which the life course is organised. This group corresponds to the conventional definition of older people and represents a period of labour market exit, reduced earnings, but without necessarily the reduction of health status and strength (Blane et al. 2004). It is expected that many people in this group will not be dependent upon their families or other institutions for support, and through social pensions may continue to be a principal breadwinner as well as caregivers to grandchildren.

9.3.5 Late Old Age (70+)

This age group is expected to be that in which the shift occurs from being a principal decision maker to a nominal decision maker, and in which the transition is also completed from bread winner to consumer. This period may also be one of extreme financial and physical dependency on families or other institutional arrangements. People in this category are thus most likely to be receiving care and least likely to be able to provide care or financial support to others.

Obviously any such categorisation is not inflexible, and there are likely to be person-by-person differences on many levels. For instance, it is possible that ill health may result in the extreme dependency of those younger than 60, while accumulated wealth may ensure that those older than 80 retain their position as principal decision maker. Furthermore, due to the heavy workload of many older people, it is quite possible that some of those in the "frail old age" category may still be performing household chores. There will also be differences between men and women. This definition of older people retains the notion of a chronological or demographic definition of ageing, but should allow for differences between social groups to emerge. We will operationalise our analysis of the life course and the near-old accordingly, and test to establish whether the distinction described in this section is found in the households sampled for NIDS.

9.4 Data Analysis

The data analysis reflected in this section uses the first wave of the South African National Income Dynamics Study that took place in 2008 in 7,302 households across the country. A full explanation of the NIDS methodology is documented by Leibbrandt et al. (2009). In the analysis that follows post-stratification weights are applied to these households and the individuals within them in order to understand the statistics at the population level. The adult sample consisted of 16,880 individuals aged 15 and above, which represents some 30 million South Africans, and in the analysis that follows, the data is weighted. In the NIDS a person is defined as an adult if they were 15 years old or older on the day of the interview, however due to inaccuracies in date of birth information there are 43 individuals who are 14 years old in the Adult file and 26 individuals who are 15 years old in the Child file (SALDRU 2009). Individuals aged under 15 have been dropped from the dataset analysed here.

Figure 9.1 shows the percentage of the population that falls within each age cohort, forming a population pyramid. As would be expected, there are slightly

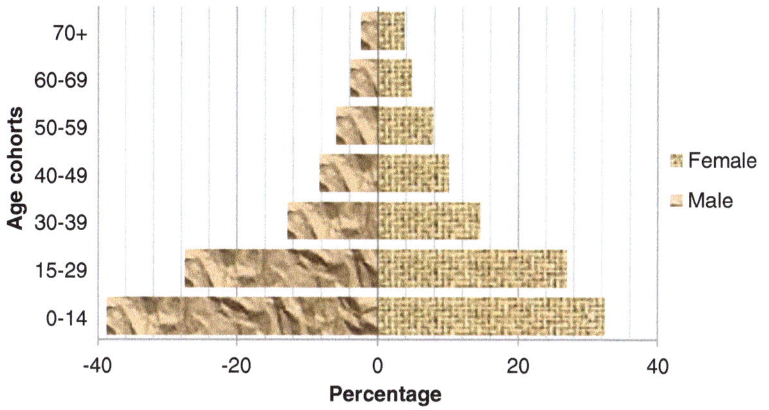

Fig. 9.1 Percentage of population by age cohort (n=25,042; N=42,563,827)

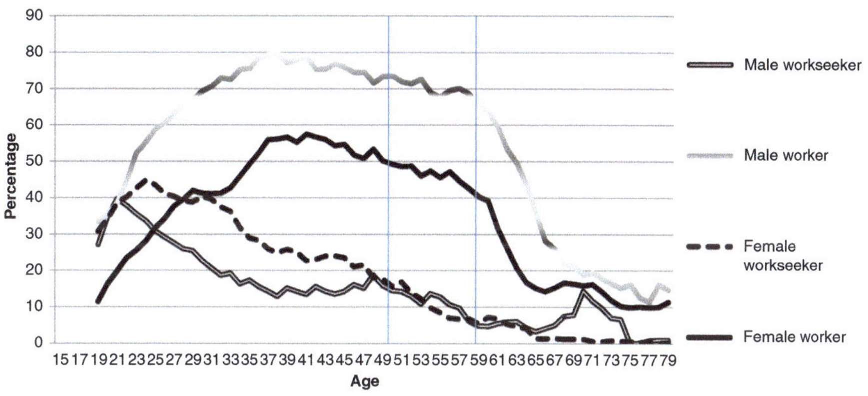

Fig. 9.2 Employment status by gender and age (n=12,712; N=22,421,965). Graph reflects moving average trendlines; results for individuals aged 80 and over have been excluded because of extremely small sample sizes or no observations at these ages

more males at younger ages and slightly more females at older ages. These percentages are in line with the mid-year population estimates generated by Statistics South Africa for 2010. Decreased life expectancy is also reflected in the shape of the population pyramid. Life expectancy for males was 53.3 years and 55.2 years for females in 2010 (Statistics South Africa 2010c). This is drastically lower than the life expectancies of 60.9 years for males and 67.7 years for females for the period 1995–2000 (Yach and Buthelezi 1995), and suggests the impact of AIDS.

Employment information from NIDS was provided by respondents and where respondents were not available or unable to answer their own adult questionnaire it was given by proxy. Actual ages and not age categories are used in order to provide more nuanced findings, although in each instance the findings for the near-old are highlighted. In Fig. 9.2 it is clear that at all ages more men than women are employed, similar numbers of women as men tend to be employed from later ages and there is a drop off in their employment that occurs before that of men's. Moreover a higher percentage of men than women want to be employed in all age groups, up to the early fifties. After this point there are, on average, more women than men seeking employment. There is an overall decline in employment from the early forties and a sharp drop-off in employment from pensionable age. The decline is steeper for women than men. The findings shown in Fig. 9.2 echo Ranchod's (2009, p. 3) point about very low employment rates in general, high unemployment at the younger ages, and labour force withdrawals at relatively early ages in South Africa: one-third of the adults aged 50–54 and 44% of adults aged 55–59 are not economically active (Ranchod's 2009, p. 3).

The findings for the near-old using the NIDS in 2008 contrast with those from the 1996 Census, over a decade before. At this time 44% of the near-old were employed and 13% unemployed (Hunter and May 2003, p. 21). The same figures from the 2008 NIDS are 54% and 9%, respectively, suggesting an overall improvement in labour market conditions. However this is notwithstanding the effects of the global recession which are unlikely to be reflected in the NIDS data which were collected between February and July in 2008.

Employment trends across race groups were also calculated for each age. The racial classification used by Statistics South Africa (2000) has been applied in this chapter. Similar trends are evident across the race groups, although for almost all ages a higher proportion of whites and coloureds are employed than Africans. After retirement age the picture changes with proportionately more Africans in employment than coloureds and whites, indicative of the fact that for many Africans retiring is not an option. This supports the idea that older Africans may withdraw from the labour market but not from work. Nevertheless there is a steady decline in the proportion of the African near-old who are employed. There is a decrease for the other race groups, though with variation due to small sample sizes. From this and other information it is clear that the life course for Africans in the South African context is in many ways unique, showing the divergence between chronological age and biographical time. Discrimination on the basis of race in apartheid era South Africa has left a permanent mark on the employment profile of a cohort of Africans, and it is the African life course that will from this point be the focus of this chapter.

In Fig. 9.3 different types of employment are reflected per age as a percentage of the employed for African males. It can be seen that a large proportion of those of working age are wage earners; there is an overall decline in the proportion of wage earners from the early forties, including for the near-old and an increase in employment for those aged 60, followed by a steady decline in wage employment. There seems to be an increase in self-employment for near-old males (suggesting that for this group this may be an option that is easier to access than wage employment), followed by another increase for males in their sixties. The percentage who undertake casual work remains relatively constant across each age, although it is highest for the youth. There is also a noteworthy increase of those in their sixties and seventies undertaking subsistence agriculture, although it should be borne in mind that those at these ages make up the smallest proportion of the population, and the implications are not that noteworthy. Finally, as we would expect, the percentage of those in receipt of a pension (this includes state, private, and foreign pension, private

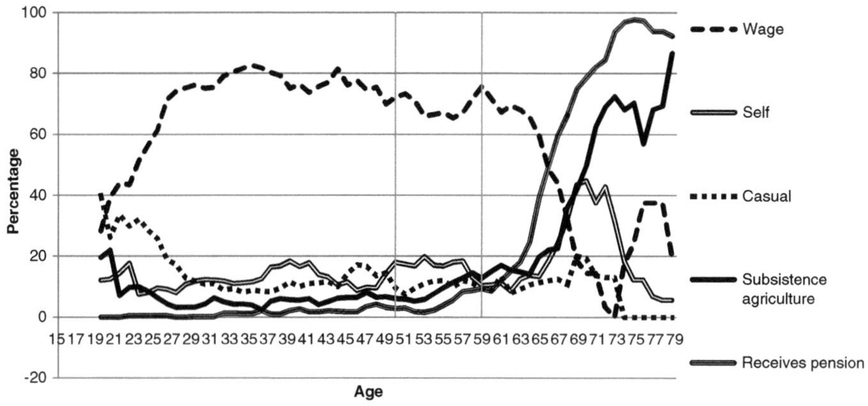

Fig. 9.3 Type of employed by age, male African only. Graph reflects moving average trendlines; results for individuals aged 80 and over have been excluded because of extremely small sample sizes or no observations at these ages; all trendlines reflect type of employment as a percentage of the employed except "receives pension" is as a percentage of each age. *n*=2,013; *N*=4,390,670 (employment); *n*=3,643; *N*=7,265,425 (pension receipt); results for individuals aged 80 and over have been excluded because of extremely small sample sizes or no observations at these ages

retirement annuity, retirement gratuity/package) at each age increases markedly from the age of about 60, although this is preceded by a slight increase in those receiving a non-state pension in their fifties. With regard to the state OAP, at the time of the first NIDS in 2008 women were eligible at 60 and men were only eligible at 65, but since 2010 the age eligibility is the same for men and women at 60 years.

The employment situation is quite similar for female Africans, as Fig. 9.4 shows, but substantially less are in wage employment and females receive a pension sooner than males. At later ages more women than men are involved in subsistence agriculture. Self-employment and casual employment remain relatively consistent in terms of percentages across the life course, which was not evidenced with males. Overall, while there seems to be an overall decline in employment for near-old African males and females, it is not as steep as expected.

In order to ascertain the trends in income by age, total individual income was calculated. In South Africa, unlike in many other sub-Saharan African countries, the majority of those aged 60 or over receive a pension of some sort. Given our focus on the near-elderly, we therefore exclude those aged sixty and over from the remainder of the analysis. This is done by adding together the same income sources that constitute total household income for the NIDS data: labour market income, government grant income, other government income, investment income, and remittance income. Income of a capital nature (such as inheritance, gift income, retrenchment payments) is excluded from the aggregate income calculation because of its distortionary effect (Argent 2009).

Figure 9.5 shows median and mean income by age for Africans—the median is a more stable measure of central tendency than the mean, but providing both

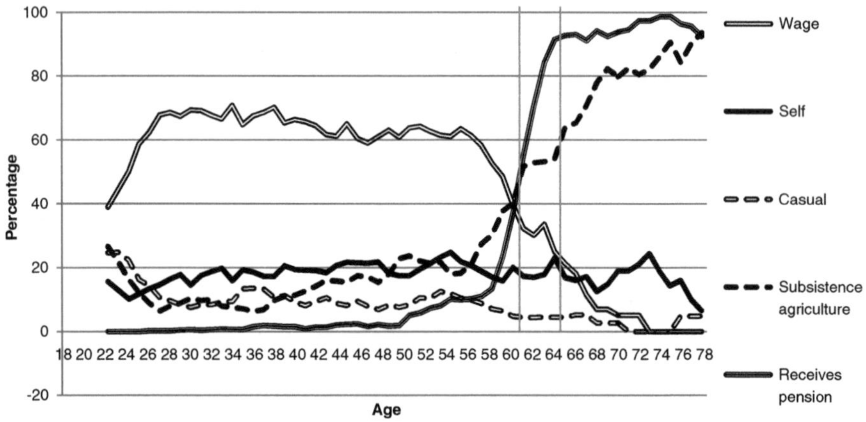

Fig. 9.4 Type of employment by age, female African only. Graph reflects moving average trendlines; results for individuals aged 80 and over have been excluded because of extremely small sample sizes or no observations at these ages; all trendlines reflect type of employment as a percentage of the employed except "receives pension" is as a percentage of each age. $n=2{,}086$; $N=3{,}730{,}311$ (employment); $n=6{,}025$; $N=9{,}933{,}896$ (pension receipt); results for individuals aged 80 and over have been excluded because of extremely small sample sizes or no observations at these ages

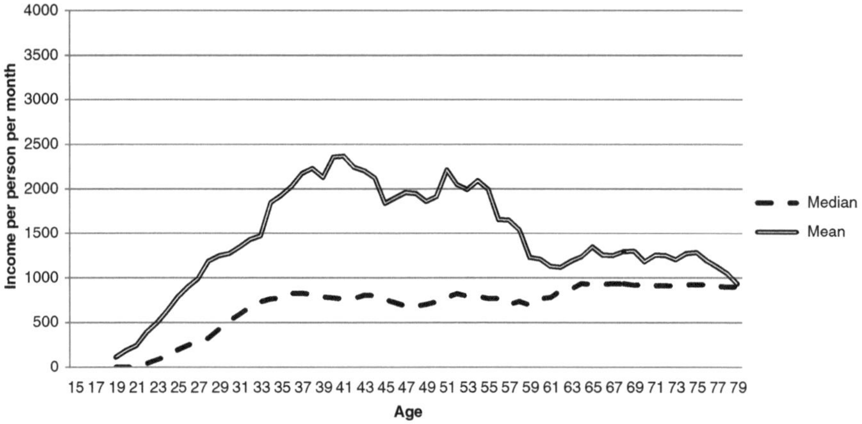

Fig. 9.5 Individual income by age, African only ($n=9{,}668$; $N=17{,}199{,}321$). Graph reflects moving average trendlines; results for individuals aged 80 and over have been excluded because of extremely small sample sizes or no observations at these ages

measures is illustrative of variance in the data. The youth have the lowest mean and median income and it is only from the mid-thirties that median income levels become relatively constant. Highest mean income levels are seen for those in their thirties and forties, and there is a gradual decrease in mean income for the near-old. Both mean and median income stabilises from the retirement ages to almost the lowest mean income and the highest median income of all the ages. If only the

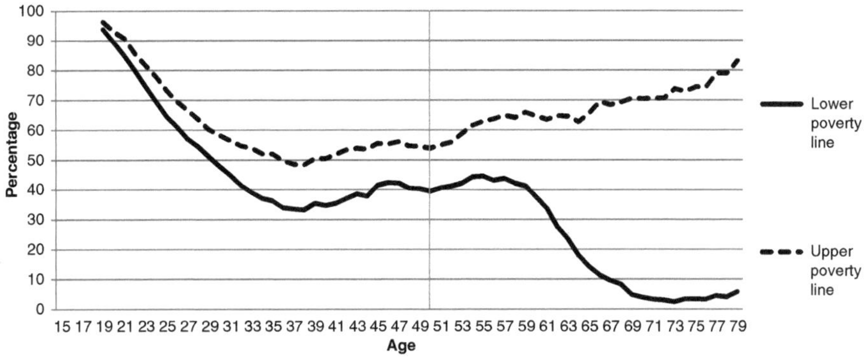

Fig. 9.6 percentage poor by age ($n=9,668$; $N=17,199,321$). Graph reflects moving average trendlines; results for individuals aged 80 and over have been excluded because of extremely small sample sizes or no observations at these ages

median income is considered, it shows that many Africans receive their highest income when they retire. This reveals a polarisation of the African work experience: for half of the population retirement brings their first taste of the good life since they have been excluded from an adequate income until their receipt of the OAP; for others, retirement brings about a decline in income. There are some factors that may contribute to a relatively high income for the near-old. Those in wage employment are likely to be earning higher wages than those at younger ages because they are likely to have been in their positions for longer. Moreover, where wage income is being lost, it may be substituted by an increase in remittance income and self-employment income, as well as an increase in subsistence agriculture.

Figure 9.6 shows the percentage of the individuals in each age cohort that are poor using two absolute poverty lines in Rand, 2008 (following Leibbrandt et al. 2010): an upper poverty line of R949 per capita per month and a lower poverty line of R515 per capita per month. These amounts are appropriate to apply to the NIDS data which was obtained in 2008. The proportion of the near-old who are poor using both measures is strikingly similar to that of the population in their thirties and forties, although slightly more of the near-old fall beneath the upper poverty line than those in these earlier decades. It is however the poverty of the youth that is striking, with very high percentages of those aged 15–29 defined as poor. What is also evident is that the state pension (R940 at the time the study was undertaken) substantially reduces extreme poverty for those at pensionable age.

The near-old are marginally poorer than those in their mid-thirties to mid-forties. The poverty of the youth also draws our attention. According to Ardington and Case (2009), the NIDS provides an ideal vehicle to understand the joint determination of economic status and health status in South Africa. The most useful measure of a person's health in the 2008 NIDS data is perceived health status. This is clearly not an ideal measure as it is dependent on the view of the person themselves and not on more rational clinical judgment. It is, nevertheless, still instructive. All adults were asked to rate their health status on a scale ranging from "excellent" to "poor".

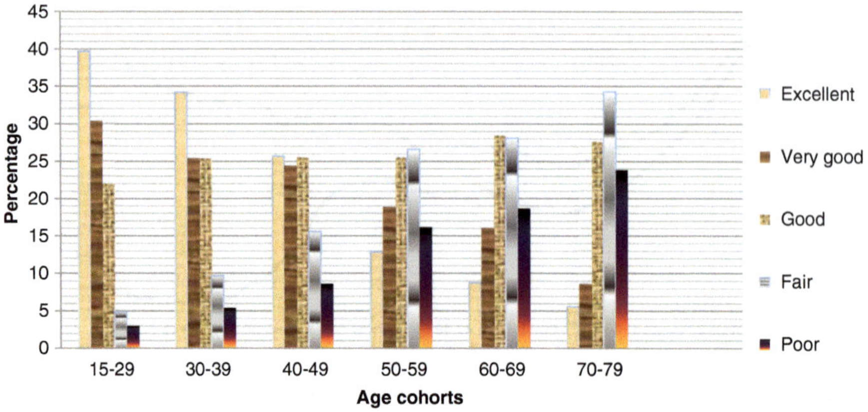

Fig. 9.7 Perceived health status by age cohort ($n=9{,}601$; $N=17{,}066{,}852$). Results for individuals aged 80 and over have been excluded because of extremely small sample sizes or no observations at these ages

For purposes of illustration, all age cohorts are included in Fig. 9.7, which shows that perceived health status becomes more "negative" with age, with more of those in younger age cohorts regarding their health to be in a good state, and more of those in older age cohorts perceiving their health to be in a less satisfactory condition. For instance, most of those in their seventies perceive their health status to be "fair" or "poor", while most of those aged 15–29 perceive their health be "excellent" or "very good", both of which are to be expected since physical health tends to deteriorate with age.

9.5 Multivariate Analyses

Thus far the analysis has been descriptive in nature. Multivariate analyses are able to take the investigation a step further and reveal causality. To begin, however, those aged 60 and over are excluded from the analysis since they are not directly comparable to individuals at younger ages who are not eligible for a state pension at minimum.

Various weighted multivariate models were constructed using as dependent variables individual income, formal employment, and "excellent" and "poor" perceived health status. These models revealed that across the life course, for those aged 15–59, the most vulnerable group in terms of employment and income, and therefore to being poor, are the youth (aged 15–29). This should not be surprising since the unemployment rates for South African youth are very high. In contrast, in terms of health, being "young" predicts an "excellent" perceived health status, while being near-old predicts a "poor" perceived health status. Therefore we also exclude the youth from the analysis that follows, in order to test our hypothesis about the near-old with regard to those aged 30–49.

Table 9.1 Linear regression model predicting total individual income (n=4,155)

Individual/household characteristics	Coef.	Std. Err.	t	$P>\|t\|$	95% conf. interval	
Female	–546.65	102.44	–5.34	0.00	–747.50	–345.81
Age 30–34	–742.48	153.13	–4.85	0.00	–1042.70	–442.26
Age 35–39	–174.55	156.27	–1.12	0.26	–480.92	131.82
Age 45–49	86.90	169.50	0.51	0.61	–245.41	419.20
Age 50–54	475.86	180.12	2.64	0.09	122.72	828.99
Age 55–59	340.55	197.85	1.72	0.09	–47.34	728.45
Highest grade	228.56	17.60	12.98	0.00	194.05	263.07
Formal employment	2164.30	107.91	20.06	0.00	1952.75	2375.86
Rural	–392.87	100.80	–3.90	0.00	–590.50	–195.24
Excellent PHS	369.05	138.28	2.67	0.01	97.95	640.16
Very good PHS	132.37	137.01	0.97	0.33	–136.24	400.99
Fair PHS	151.84	157.95	0.96	0.34	–157.82	461.50
Poor PHS	166.84	187.03	0.89	0.37	–199.83	533.52
Constant	–531.35	223.92	–2.37	0.02	–970.35	–92.35

PHS perceived health status
Prob > F = 0.0000
Adj R-squared = 0.1878

Table 9.1 shows a weighted linear regression model with total individual income as the dependent variable. The model explains 19% of the variance in the outcome variable. Being female and being 30–34 years of age are negative predictors of income, while being 50–54, having an excellent perceived health status, the highest school grade, being in formal employment, and living in a rural area are all positive predictors of individual income. Therefore, being 30–34 is one of the determinants of lower income, while being near-old is a determinant of higher income.

A logistic regression model (see Table 9.2) with formal employment as the dependent variable, and which explains 12% of the variance in the outcome variable, shows that being female, being 30–34, having a "poor" perceived health status and living in a rural area are all negative predictors of formal employment. Having formal employment is defined as being paid a wage or salary to work on a regular basis for an employer, either full-time or part-time. The highest school grade and having an "excellent" perceived health status are the positive predictors of formal employment. Therefore, being near-old does not determine formal employment, rather being in one's early thirties does.

Table 9.2 showed that perceived health status seems to be linked to age, with more of those in younger age brackets defining their health status positively and more of those in older age brackets classing their health status negatively. The logistic regression models that follow attempt to understand if there is a causal link between the two, that is, if being near-old means that one will have worse health than if one is not near-old. The regression model given in Table 9.3 explains only 7% of the variance in the outcome variable. A number of factors predict having an "excellent" perceived health status. Positive predictors include

Table 9.2 Logistic Regression model predicting formal employment ($n = 4{,}155$)

Individual/household characteristics	Odds ratio	Robust Std. Err.	z	$P>\|z\|$	95% conf. interval	
Female	0.377	0.04	–9.88	0.00	0.31	0.46
Age 30–34	0.68	0.10	–2.53	0.01	0.50	0.92
Age 35–39	0.81	0.12	–1.35	0.18	0.60	1.10
Age 45–49	0.99	0.16	–0.04	0.97	0.72	1.37
Age 50–54	1.01	0.18	0.06	0.95	0.71	1.44
Age 55–59	1.17	0.23	0.82	0.42	0.80	1.73
Highest grade	1.18	0.02	9.07	0.00	1.14	1.23
Rural	0.49	0.05	–7.09	0.00	0.40	0.60
Excellent PHS	1.33	0.17	2.15	0.03	1.03	1.72
Very good PHS	1.03	0.14	0.23	0.82	0.79	1.35
Fair PHS	0.79	0.13	–1.47	0.14	0.58	1.08
Poor PHS	0.37	0.09	–4.18	0.00	0.23	0.59

PHS perceived health status
Wald chi-squared = 290.43
Prob > chi-squared = 0.0000
Pseudo R-squared = 0.1205

Table 9.3 Logistic regression model predicting "excellent" perceived health status ($n = 3{,}620$)

Individual/household characteristics	Odds ratio	Robust Std. Err.	z	$P>\|z\|$	95% conf. Interval	
Male	1.37	0.16	2.75	0.01	1.09	1.72
Age 30–34	1.77	0.34	3.00	0.00	1.22	2.58
Age 35–39	1.64	0.31	2.66	0.01	1.14	2.37
Age 40–44	1.63	0.31	2.52	0.01	1.11	2.38
Age 50–54	0.72	0.17	–1.37	0.17	0.45	1.15
Age 55–59	0.41	0.11	–3.19	0.00	0.24	0.71
Married	1.22	0.14	1.70	0.09	0.97	1.54
Highest grade	1.07	0.02	3.23	0.00	1.03	1.12
Formal employment	1.40	0.18	2.65	0.01	1.09	1.80
Individual income	1.00	0.00	–0.65	0.51	1.00	1.00
Rural	1.77	0.28	3.59	0.00	1.30	2.43
Piped water	1.27	0.18	1.66	0.10	0.96	1.68
Flush toilet	2.09	0.38	4.09	0.00	1.47	2.98
Electricity	0.78	0.11	–1.73	0.08	0.59	1.03

Wald chi-squared = 138.78
Prob > chi-squared = 0.0000
Pseudo R-squared = 0.0690

being male, being in the thirties and early forties, the highest grade achieved, having formal employment, living in a rural area, and having a flush toilet—in fact the odds of having an "excellent" perceived health status is over two times greater if a person has access to a flush toilet. The only negative predictor is being age 55–59.

Table 9.4 Logistic regression model predicting "poor" perceived health status (n=3,620)

Individual/household characteristics	Odds ratio	Robust Std. Err.	z	$P>\|z\|$	95% conf. interval	
Female	1.41	0.26	1.87	0.06	0.98	2.02
Age 30–34	0.52	0.17	−1.99	0.05	0.27	0.99
Age 35–39	1.26	0.34	0.85	0.40	0.74	2.13
Age 45–49	1.43	0.38	1.35	0.18	0.85	2.42
Age 50–54	1.82	0.50	2.19	0.03	1.06	3.11
Age 55–59	2.08	0.59	2.58	0.01	1.19	3.65
Married	0.74	0.12	−1.86	0.06	0.55	1.02
Highest grade	0.89	0.02	−4.33	0.00	0.85	0.94
Formal employment	0.37	0.10	−3.80	0.00	0.23	0.62
Individual income	1.00	0.00	−1.64	0.10	1.00	1.00
Rural	0.58	0.14	−2.24	0.03	0.36	0.93
Piped water	0.78	0.16	−1.20	0.23	0.53	1.17
Flush toilet	0.75	0.20	−1.07	0.29	0.44	1.28
Electricity	1.14	0.24	0.61	0.55	0.75	1.71

Wald chi-squared(14) = 152.35
Prob > chi-squared = 0.0000
Pseudo R-squared = 0.1101

The regression model, given in Table 9.4 which has "poor" perceived health status as the dependent variable, explains 11% of the variance in the outcome variable. This time being 50–54 and being 55–59 years old are positive predictors of "poor" perceived health status. Negative predictors are being age 30–34, the highest grade achieved, being formally employed, and living in a rural area. In many ways this is an inverse of the previous model.

Therefore multivariate analyses show that being near-old does not predict a decline in income nor employment—in fact it predicts the likelihood of higher income—but being near-old does predict a poor perceived health status controlling for income, access to services, sex, and employment status.

9.6 Discussion

In this chapter we have focused on the near-old, since we have hypothesised that in the African context the vulnerability associated with old age is more likely to occur at younger ages than at the commonly used definition of retirement age. The chapter has used a life course framework to assess the employment, income, and health situation of the near-old in South Africa. In line with Hunter and May (2003) it was our expectation that in terms of the labour market, income, and health, interesting and worrying findings would arise with regard to this age group. Although the near-old may be vulnerable both economically and in terms of their health, our findings from the NIDS data show that once education, location, and sex have been controlled for, this group is neither more likely to be without formal employment nor to have lower incomes than those in younger age groups. Indeed, our analysis confirms the suggestion made by Lombard and Kruger (2009) that older people remain in the labour force, and the fact that Africa is the region with the highest labour force participation of older workers (ILO 2010a). It is the youth who are worse off with regard to employment and income when compared to the near-old. The ILO (2010a) points out that when compared to other age groups, young people present low and declining participation rates, and the highest unemployment. This is of all the more concern in South Africa since the youth make up such a large portion of the population. However, in terms of their health and well-being, the near-old are more vulnerable than those at younger ages. Therefore the proposed vulnerability of the near-old is not confirmed with regard to income and employment, but it is established in relation to health status.

There are various reasons why the situation of the near-old differs from our earlier analysis which did suggest greater vulnerability among this group (Hunter and May 2003). Firstly, at least some of those previously identified as vulnerable have now aged into the pensionable age groups. Further, the overall employment situation of the near-old has improved since 1996. In addition, those who are currently near-old progressed from youth through middle age at a time of comparative full employment, and even if not presently employed, are likely to have experienced some employment over their life course. In contrast, the youth today are growing up in a context of widespread unemployment and may not experience employment. Finally, it is likely that any loss of formal employment by the near-old is offset by other livelihoods (possibly the start of entrepreneurial activity), or the consumption of savings, or alternatively it may be their preference to go out of the labour market.

In many ways South Africa is unique when compared to other countries in Africa, and this is no different with regard to the situation of older people. Because of its apartheid past, white older people can be distinguished from African older people in various ways. As Statistics South Africa (2010b, p. 100) outline, "the relatively high proportion of whites that survive to old age to some extent masks the desperate plight of particularly the African elderly when looking at averages for the group as a whole". There is a substantial difference between the two race groups with regard to the ageing index—that is, the number of people aged 65 and over per 100 youths under age 15. For Africans the figure is 12, while for whites it is 77 (Day and Gray 2010, p. 222). Almost 17.9% of the white population is aged 60 and over, compared with only 6.6% of the African population (Day and Gray 2010, p. 223). Moreover, projected life expectancy at birth for 2009 was 47.3 years for Africans and 71.5 years for Whites (South African Institute for Race Relations 2010, p. 52). There is no doubt that the HIV/AIDS pandemic is largely responsible for this difference. Among the African population it is to a large extent those who do not die of AIDS at younger ages who live to older ages.

What then are the implications of living longer for those Africans who do make it to older ages? They are likely to be dependent on the OAP for their own but also their family's well-being, because of a lack of or inability to save for old age. They are likely to still work to supplement their pension income, if their health enables this. Their health is likely to be in decline compared to those at younger ages.

Those with access to additional forms of income may be experiencing better health than their counterparts who are not.

South Africa also differs from the developed world in another aspect that relates to older people. While the eligible age for men for the state OAP was reduced from 65 to 60 (in line with age eligibility for women), in the developed world the age eligibility is being extended so that older people are only able to access pensions at later ages which means they are working longer. This change in age eligibility is because of a Constitutional challenge in favour of men: a number of applicants argued that the provisions discriminated against them unfairly on the grounds of age and gender, and contradicted the equality clause in the Constitution (Lombard and Kruger 2009). It seems unlikely that the eligible age for the OAP will be extended in the future.

The life course framework has proved useful in revealing vulnerable groups. While we hypothesised that the near-old would be one of the most vulnerable, it seems that it is in fact the youth who are most at risk in South Africa. Africans aged 60 and over are not the most vulnerable because of the important role played by the OAP. Youth unemployment is also a challenge in developed countries (see O'Higgins 1997), particularly since the global financial crises where the youth have fared worst in terms of unemployment (ILO 2010b). In South Africa youth unemployment can be attributed in part to a lack of work experience (with regard to wage employment) and a lack of access to capital (with regard to self-employment)—something which those at older ages have a higher likelihood of accumulating over their lifetimes (Mlatsheni and Rospabé 2002). A focus on the youth would seem most urgent considering the scale of their employment and income problem. This will require an approach that focuses on the micro- and macro-economic levels, as well as labour demand and supply (ILO 2010a). Although the situation of the current cohort of the near-old appears to be an improvement upon those that have recently graduated from this stage of the life course, there are still important policy implications. One issue to probe may be whether policy makers are doing enough for older entrepreneurs and subsistence farmers. Secondly, unless this group is simultaneously saving towards their own retirement, they are likely to face a sharp decline in their incomes upon their retirement. Unless they are able to adjust their expenditure patterns, or find alternative sources of income, retirement may now bring hardship rather than relief from poverty. Finally, it may also be useful to explore other dimensions of the life course, such as widowhood.

References

Alavinia, S. M., & Burdorf, A. (2008). Unemployment and retirement and ill-health: a cross-sectional analysis across European countries. *International Archives of Occupational and Environmental Health, 82*, 39–45.

Aliber, M. (2001). *Study of the incidence and nature of chronic poverty and development policy in South Africa: An overview. Report*. South Africa: School of Government, University of the Western Cape.

Ardington, C., & Case, A. (2009). Health: Analysis of the NIDS wave 1 dataset. Discussion Paper no. 2. University of Cape Town and Princeton University.

Ardington, E., & Lund, F. (1995). Pensions and development: social security as complementary to programmes of reconstruction and development. *Development Southern Africa, 12*(4), 557–577.

Argent, J. (2009). *National Income Dynamics Study (NIDS). Household income: Report on NIDS wave 1. Technical Paper No. 3*. Cape Town: South African Labour and Development Research Unit, University of Cape Town.

Barrientos, A. (2003). What is the impact of non-contributory pensions on poverty? Estimates from Brazil and South Africa. Unpublished paper. Institute for Development Policy and Management, University of Manchester.

Barrientos, A., Gorman, M., & Heslop, A. (2003). Old age poverty in developing countries: contributions and dependence in later life. *World Development, 31*(3), 555–570.

Blane, D., Higgs, P., Hyde, M., & Wiggins, R. D. (2004). Life course influences on quality of life in early old age. *Social Science and Medicine, 58*, 2171–2179.

Burns, J., Keswell, M., & Leibbrandt, M. (2005). Social assistance, gender and the aged in South Africa. *Feminist economics, 11*(2), 103–115.

Carter, M., & May, J. (1999). Poverty, livelihood and class in rural South Africa. *World Development, 27*(1), 1–20.

Carter, M., & May, J. (2001). One kind of freedom: poverty dynamics in post-apartheid South Africa. *World Development, 29*(12), 198–206.

Chen, S., & Ravallion, M. (2008). The developing world is poorer than we thought, but no less successful in the fight against poverty. World Bank Policy Research Working Paper Series. Resource document. http://papers.ssrn.com/sol3/papers.cfm?abstract_id=1259575. Accessed 23 Feb 2011.

Chen, Y., Wadensjo, E., & Tull, A. (2009). Potential labor supply and flexible work options for all workers: an exploratory essay. *European Papers on the New Welfare, 11*, 49–57.

Choudhury, S., & Leonesio, M. V. (1997). Life-cycle aspects of poverty among older women. *Social Security Bulletin, 60*(2), 17–36.

Crosnoe, R., & Elder, G. H. (2002). Successful adaptation in the later years: a life course approach to aging. *Social Psychology Quarterly, 65*(4), 309–328.

Day, C., & Gray, A. (2010). Health and related indicators. In S. Fonn & A. Padarath (Eds.), *South African health review 2010* (pp. 211–364). Durban: Health Systems Trust.

Duflo, E. (2000). Grandmothers and granddaughters: old age pension and intra-household allocation in South Africa. Resource document. Massachusetts Institute of Technology. http://econ-www.mit.edu/files/732. Accessed 8 Mar 2011.

Elder, G. H. (1975). Age differentiation and the life course. *Annual Review of Sociology, 1*, 165–190.

Encel, S., & Studencki, H. (2004). Older workers: can they succeed in the job market? *Australasian Journal on Ageing, 23*(1), 33–37.

Facchini, C. (2009). The transition to retirement: a problem or a resource? *European Papers on the New Welfare, 11*, 26–39.

Ferreira, M. (2000). *Growing old in the new South Africa* (pp. 32–46). Spring: Ageing International.

Ferreira, M., Møller, V., Prinsloo, F. R., & Gillis, L. S. (1992). *Multi-dimensional survey of elderly South Africans, 1990–1991: Key findings. Monograph No. 1*. Pretoria and Cape Town: Human Sciences Research Council and HSRC/UCT Centre for Gerontology, University of Cape Town.

Friedman, J., Goodkind, D., Cuong, B. T., & Anh, T. S. (2001). Work and retirement among the elderly in Vietnam. *Research on Aging, 23*(2), 209–232.

Giele, J. Z., & Elder, G. H. (Eds.). (1998). *Methods of life course research: Qualitative and quantitative approaches*. Thousand Oaks, CA: Sage.

HelpAge International. (1996). *Assessment of the needs of older people in Shihimu, Northern Province, South Africa*. South Africa: Elim Care Group Project and Elim-Hlanganani Society for the Care of the Aged.

Heslop, A., & Gorman, M. (2002). *Chronic poverty and older people in the developing world. CPRC Working Paper No. 10.* Manchester: Chronic Poverty Research Centre.

Hungerford, T. L. (2001). The economic consequences of widowhood on elderly women in the United States and Germany. *The Gerontologist, 41*(1), 103–110.

Hunter, N., & May, J. (2003). "Growing old gracefully?": Ageing in post-apartheid South Africa. Paper presented at the Fourth African Population Conference, "Population and poverty in Africa: facing the 21st century challenges", Tunis, 8–12 Dec.

Hunter, N., May, J., & Padayachee, V. (2003). *Lessons for PRSP from poverty reduction strategies in South Africa. Working Paper No. 39.* Durban: School of Development Studies, University of Natal.

International Labour Organization (ILO). (2010a). *Employment and social protection in the new demographic context.* Geneva: International Labour Office.

International Labour Organization (ILO). (2010b). *Global employment trends for youth employment. Special issue on the impact of the global economic crisis on youth.* Geneva: International Labour Office.

Jackson, R., & Howe, N. (2009). The graying of the great powers demography and geopolitics in the 21st Century. Book summary. *European Papers on the New Welfare, 11*, 123–136.

Kimmel, D. C. (1974). *Adulthood and aging: An interdisciplinary, developmental view.* New York: Wiley.

Kingdon, G. G., & Knight, J. (2004). Unemployment in South Africa: the nature of the beast. *World Development, 32*(3), 391–408.

Kingdon, G., & Knight, J. (2007). Unemployment in South Africa, 1995–2003: causes, problems and policies. *Journal of African Economies, 16*(5), 813–848.

Kinsella, K., & Velkoff, V. A. (2001). *An aging world. U.S. Census Bureau, Series P95/01-1.* Washington, DC: U.S. Government Printing Office.

Leibbrandt, M., & Woolard, I. (2001). The labour market and household income inequality in South Africa: existing evidence and new panel data. *Journal of International Development, 13*(6), 671–689.

Leibbrandt, M., Woolard, I., & de Villiers, L. (2009). *Methodology: Report on NIDS Wave 1. Technical Paper no. 1.* Cape Town: South African Labour and Development Research Unit, University of Cape Town.

Leibbrandt, M., Woolard, I., Finn, A., & Argent, J. (2010). *Trends in South African income distribution and poverty since the fall of apartheid. OECD Social, Employment and Migration Working Papers, No. 101.* Paris: OECD Publishing.

Lombard, A., & Kruger, E. (2009). Older persons: the case of South Africa. *Aging International, 34*, 119–135.

Loretto, W., & White, P. (2006). Work, more work and retirement: older workers' perspectives. *Social Policy and Society, 5*(4), 495–506.

Lund, F. (1993). State social benefits in South Africa. *International Social Security Review, 46*(1), 5–25.

Mehta, K. (1999). Intergenerational exchanges: qualitative evidence from Singapore. *Southeast Asian Journal of Social Science, 27*(2), 111–122.

Mehta, K., & Cheung, P. P. L. (1997). *Retirement: Cross cultural perceptions and preparation by near-old in Singapore. Elderly in Asia Research Report Series. No. 97-39.* Ann Arbor: Population Studies Center, University of Michigan.

Mlatsheni, C., & Rospabé. (2002). *Why is youth unemployment so high and unequally spread in South Africa? Working Paper 02/65.* Cape Town: Development Policy Research Unit, University of Cape Town.

O'Higgins, N. (1997). The challenge of youth unemployment. *International Social Security Review, 50*(4), 63–93.

Platman, K., & Taylor, P. (2004). Introduction: themed section on age employment and policy. *Social Policy and Society, 3*, 143–144.

Powell-Griner, E., Bolen, J., & Bland, S. (1999). Health care coverage and use of preventive services among the near elderly in the United States. *American Journal of Public Health, 89*(6), 882–886.

Price, K., Alde, P., Provis, C., Harris, R., & Stack, S. (2004). What hinders and what helps? Searching for solutions to mature aged unemployment and the residential aged care workforce crisis. *Australasian Journal on Ageing, 23*(4), 177–183.
Ranchod, V. (2009). *National Income Dynamics Study (NIDS). Labour market: Analysis of the NIDS wave 1 dataset*. Cape Town: South African Labour and Development Research Unit (SALDRU), University of Cape Town.
Raymo, J. M., & Cornman, J. C. (1997). *Labor force status transitions across the life course in Taiwan, Singapore, Thailand and the Philippines: 1970–1990. Elderly in Asia Research Report Series. No. 97-44*. Ann Arbor: Population Studies Center, University of Michigan.
Sagner, A. (2000). Ageing and social policy in South Africa: historical perspectives with particular reference to the Eastern Cape. *Journal of Southern African Studies, 26*(3), 325–553.
Schofield, D. J., Passey, M. E., Earnest, A., Gloor, I. C., & Shrestha, R. (2007). Are we getting healthier as we grow older?: implications for babyboomer labor force participation. *Annals of the New York Academy of Sciences, 1114*(1), 230–240.
Shaw, L. B., & Lee, S. (2005). Growing old in the US: gender and income adequacy. *Feminist Economics, 11*(2), 174–186.
Smeeding, T. M., & Sandström, S. (2005). Poverty and income maintenance in old age: a cross-national view of low-income older women. *Feminist Economics, 11*(2), 163–174.
South African Institute for Race Relations (2010). South Africa Survey 2009/10. Demographics online. http://www.sairr.org.za/services/publications/south-africa-survey/south-africa-survey-online-2009-2010/the-economy. Accessed 20 Jul 2011.
South African Labour and Development Research Unit (SALDRU). (2009). *National Income Dynamics Study (NIDS). Wave 1. Introduction to NIDS data*. Cape Town: University of Cape Town.
Statistics South Africa. (2000). *Measuring poverty in South Africa*. Pretoria: Statistics South Africa.
Statistics South Africa. (2003). *Census 2001. Census in brief*. Pretoria: Statistics South Africa.
Statistics South Africa. (2010a). *Quarterly Labour Force Survey. Quarter 3, 2010. Statistical release P0211*. Pretoria: Statistics South Africa.
Statistics South Africa. (2010b). *Social profile of South Africa, 2002–2009*. Pretoria: Statistics South Africa.
Statistics South Africa. (2010c). *Statistical release P0302. Mid-year population estimates*. Pretoria: Statistics South Africa.
Taylor, P. (2004). Age and work: international perspectives. *Social Policy and Society, 3*(2), 163–170.
United Nations Population Division (2008). World population prospects: The 2008 revision. Population Division of the Department of Economic and Social Affairs of the United Nations Secretariat. Resource document. http://esa.un.org/unpp/p2k0data.asp. Accessed 5 Feb 2011.
Vartanian, T. P., & McNamara, J. M. (2002). Older women in poverty: the impact of midlife factors. *Journal of Marriage and Family, 64*(2), 532–548.
Walker, A. (1999). The future of pensions and retirement in Europe: towards productive ageing. *The Geneva Papers on Risk and Insurance, 24*(4), 448–460.
Wilson, G. (2000). *Understanding old age: Critical and global perspectives*. London: Sage.
World Health Organization (2002). Information needs for research, policy and action on ageing and older adults: Report of a workshop on creating a minimum data set (MDS) for research policy and action on ageing and the aged in Africa. http://www.who.int/healthinfo/survey/ageing_mds_report_en_harare.pdf. Accessed 23 Oct 2011.
World Health Organization (2011). Definition of an older or elderly person: Proposed working definition of an older person in Africa for the MDS Project. http://www.who.int/healthinfo/survey/ageingdefnolder/en/index.html. Accessed 23 Oct 2011.
Xu, K., Evans, D. B., Kawabata, K., Zeramdini, R., Klavus, J., & Murray, C. J. L. (2003). Household catastrophic health expenditure: a multicountry analysis. *Lancet, 362*(9378), 111–117.
Yach, D., & Buthelezi, G. (1995). Health status. In D. Harrison (Ed.), *South African health review 1995*. Durban: Health Systems Trust.

Chapter 10
Understanding the Experiences of the Elderly in Rural Areas in Rwanda

Ganzamungu Zihindula and Pranitha Maharaj

10.1 Introduction

Rwanda is a landlocked country of approximately 26,338 km^2 and is situated in east-central Africa. With a population of just over 11 million, Rwanda remains one of the most densely populated areas in Africa. The population is young and predominantly rural. More than 80% of the population lives in rural areas, engaging in subsistence agriculture (NISR and Macro International 2005). The population consists of three main ethnic groups: Tutsi, Hutu, and Twa. The genocide of 1994 led to the mass slaughter of almost 20% of the population of Rwanda and large numbers of refugees and displaced persons. The genocide was sparked by the death of the Rwandan President Juvenal Habyarimana, a Hutu, when his plane was shot down as it prepared to land at Kigali airport on 6 April 1994. Soon after the announcement of the plane crash, violence erupted around the whole country. The majority of Hutus, which constituted approximately 85% of the population, turned on the Tutsis who were only about 12% of the population and moderate Hutus, killing an estimated total of 800,000 people (Madsen 1999; Prunier 1995).

The country continues to suffer from the aftermath of the genocide. At present, the country is ranked 152 out of 169 countries on the human development index (UNDP 2010). Poverty is widespread, with the 2010 Human Development Report indicating that 77% of the population lives on less than $1.25 a day (UNDP 2010). The incidence of poverty is higher in rural areas than urban areas (Sangano et al. 2003). In the past few decades rural areas in the country have had to deal with a number of shocks in addition to the genocide. Rwanda is relatively tiny with over 350 inhabitants per square metre which has contributed to some of the problems plaguing the country. Adding to the stress is recurrent droughts and unpredictable

G. Zihindula (✉) • P. Maharaj
School of Built Environment and Development Studies,
University of KwaZulu-Natal, Durban, South Africa
e-mail: tzihindula@AfricaCentre.ac.za; Maharajp7@ukzn.ac.za

P. Maharaj (ed.), *Aging and Health in Africa*, International Perspectives on Aging 4,
DOI 10.1007/978-1-4419-8357-2_10,

climatic conditions in the past few years which have increased the pressure on the land. Most Rwandans suffer from diseases that could be relatively easily prevented and/or cured through improved access to safe water and sanitation. Malaria, tuberculosis, and dysentery, together with HIV/AIDS, have been identified as health problems directly impacting rural households, affecting working age adults as well as children and the elderly (Donovan et al. 2003). The high burden of health problems continues to threaten the country's economic and social progress, and also undermines any attempts at redressing poverty. According to the 2005 demographic and health survey, 3% of the Rwandan population aged 15–49 is infected with HIV (3.6% for women and 2.3% for men) (NISR and Macro International 2005).

The number of older men and women in Rwanda has been steadily increasing and most (64%) of the ageing population reside in rural areas. Women constitute a larger share of the older population than men. Some estimates suggest that women make up 55% of the population aged 55 and above and men are 45% (Sangano et al. 2003). Projections suggest that the ageing population will continue to increase in the next few decades. This is supported by ageing data for the period 1997–2002. In 1997 the number of older people was 428,300 and increased to 627,427 in 2002 (Sangano et al. 2003).

Most scholars who have studied ageing in relation to poverty and vulnerability express the need for some form of support for the elderly within communities. This understanding has led to the conclusion that the elderly in most instances are among the most vulnerable groups in society. In India, Murthy et al. (2005) found that the vast majority of older people in underserved areas of the country were malnourished, lived with a weak immunity, and were prone to a wide range of infectious and chronic diseases. Barrientos et al. (2003, p. 558) observe that "poverty tends to take a 'U' shape, that is; it is higher among the younger and the older members of the society." They attribute the high levels of poverty among the elderly to the absence of income, inadequate family and social support, as well as poor health. Moreover, the absence of pension plans is also a central cause of poverty in later life and this often forces the elderly to continue working after their retirement age, although job opportunities for them remain scarce (Barrientos et al. 2003, p. 558). In the rural areas the elderly depend heavily on households and social networks for their livelihoods. In Asia, Barrientos et al. (2003) note that the elderly with no family support often obtain informal assistance from neighbours.

Hampson (1990) argues that traditionally, the elderly were accorded an elevated status in society and were highly respected because they were considered to be spiritual or religious leaders. According to Rwezarura (1989, p. 5), "the older the individual became the higher were his chances of gaining upward mobility in the social hierarchy." Prior to the colonial era, the elderly had economic power as they were the ones who controlled the means of production, i.e. the land, livestock, and harvests (Hampson 1990). With the advent of colonisation and industrialisation, Africans lost their land, and agriculture was no longer the core economic activity, hence the elderly lost control of the means of production. According to Hampson (1990), this change led to a shift in the perception of and the authority held by the elderly. Hampson (1990, p. 16) aptly describes their situation arguing that

> The elderly, even after independence remained marginalised, and this has left them pauperised and socially excluded. They are also not a priority in government and even international organisational planning. The elderly are impoverished because most of them are not educated, which means that they cannot get access to post retirement employment as education is now a prerequisite to obtain a job. They had previously been viewed as embodying the value systems of the society and played the role of educating the younger generation on matters of cultural value. Today however, very little appreciation is being given to the role of cultural educator which the elderly previously held.

Rwezarura (1989) also writes about the changing roles and obligations of the elderly in Africa. Traditionally, the elders were generally in control and had influence on a number of issues just because they had the authority to control strategic resources such as land and livestock. However, attitudes towards the elderly are changing. In many parts of Africa, the introduction of the monetary economy and new forms of social and political control has drastically weakened the position of the elderly, resulting in the loss of their authority, power, privilege, and leadership positions in the community (Rwezarura 1989). Hampson (1990) argues that the impact is particularly severe for women. Elderly women, especially widows, are more likely than elderly men to live alone, more likely not to have livestock, more likely not to be in contact with children and close members of the family, and are more vulnerable financially, socially and culturally.

To date there has been relatively limited research on the elderly living in rural areas. This chapter attempts to fill this gap by examining the critical issues facing older people in rural areas in Rwanda. The specific aim of the study was to deepen understanding of the experiences of older people in the rural areas in a country that is emerging from conflict. Research suggests significant rural–urban differences in health, income, and availability of services (Penning and Chappell 1994). Krout (1986) argues that rural areas are different from urban areas due to population size and density and geographic, economic, social, and cultural factors. Further findings suggest that, in general, rural areas are characterised by greater poverty, less adequate housing and transportation systems, poorer health, greater incidence of chronic health conditions, and lack of a wide range of services (Bull and Bane 1993; Krout 1994). According to Krout (1986), rural residents present basic differences in needs and resources compared with urban residents.

10.2 Study Site

The study was conducted in two rural areas of Rwanda, namely, Rwamagana and Nsinda village. The former area is located 202 kilometres from the capital city, and the latter is almost 156 kilometres on the west side of the capital city. These study sites were selected because they display many of the characteristics of rural areas. Most parts of the areas are not readily accessible by public transport and most households do not have access to electricity or safe drinking water. Although Rwanda is one of the most densely populated regions in Africa, the population in rural areas are scattered on hilly terrain, which is very difficult to access. Observing the area from afar, the population is widely dispersed, with almost 1–2 km separating each dwelling, unlike in urban areas where houses are in close proximity to each other.

10.3 Methodology

A qualitative approach was identified as the most appropriate method to understand the experiences of older people in rural areas. Qualitative data provide in-depth descriptions and reveal individuals' understandings of events or actions, which can help the researcher to gain insight into why and how those events or actions take place rather than just representing a phenomenon (Babbie and Mouton 2001). Qualitative research methods also allow respondents to express feelings and opinions in their own words. In this study, open-ended interviews were employed as the principal data collection tool. An attempt was made to ensure an equal distribution of men and women. In total, 20 in-depth interviews were held: 10 men and 10 women. The youngest respondent was a male aged 60 years, and the oldest was a female aged 91 years. Only nine of the respondents reported having ever attended school. Level of education was very low with only one having completed secondary school. Most were dependent on their land for their livelihoods. Agriculture was the main source of income. All the respondents had practiced subsistence farming and for those who are no longer strong enough to cultivate the land, it is being done by their children, grandchildren, or other family members. Although none of the respondents were employed at the time of the interviews, four reported that they had been working in the past but had since retired. These four were surviving on the pension that they received from the government. However, they described their life as difficult despite receiving grants and other assistance from some organisations. All the respondents had been married at one time or another in their lives. At the time of the interviews, 12 respondents reported that they had lost their partners due to unforeseen circumstances and they were widowed. The remaining 8 were still married to their partners, but some were experiencing marital problems.

The in-depth interviews collected information on a range of topics including: experiences of living in a rural area, access to health facilities, challenges respondents face in coping with their socio-economic situation, and their perceptions as well as their understanding of their situation in general. Prior to the interviews, the village chief was approached and informed about the study. The chiefs were not able to assist in recruiting respondents for the study but they granted the researchers permission to go door to door to secure interviews, accompanied by two locals. Some elderly welcomed the opportunity to participate in the study and others refused. They remained suspicious of the motives of the researcher. The interviews took place at the residence of the respondent. All attempts were made to ensure maximum privacy during the interviews. Interviews were recorded with the permission of each respondent but some refused because they were afraid of the consequences. Each respondent was assured that there were no negative comebacks but it was difficult to gain his or her complete trust. In some cases, only notes were taken during the interview.

10.4 Analysis

As with any type of data analysis, there exist various methods for making sense of the data. This study adopted and identified thematic analysis as the most appropriate method of analysis. Thematic analysis identifies common ideas and phrases that individuals articulate in their narratives and that can indicate some degree of importance allocated to a specific thought or occurrence. This study draws on the method of thematic analysis utilised by Owen (1984) which considers three aspects for identifying themes. The first criterion is recurrence, which refers to concepts that are repeated using similar phrases or words. The second term is repetition, meaning that an idea is conveyed with the use of the same words. And the third is forcefulness, which refers to the emphasis applied to a concept.

The identification of themes or similarities in the narrative data occurs at two levels, intra-thematic and inter-thematic. Intra-thematic analysis considers the themes that are identified for that particular respondent. Inter-thematic analysis compares themes among respondents. Using methods of thematic analysis on both levels allows for the data to be thoroughly reviewed and compared in a systematic, reproducible manner (Owen 1984). Data is often transcribed and coded to enhance thematic identification. Therefore, these steps were followed fully during data analysis.

10.5 Findings

There are various factors that affect the decision of the elderly to live in rural areas. Among the main reasons encouraging the elderly to live in the rural areas was socio-economic factors. The elderly are often forced to move to the rural areas because of the high cost of living in the urban areas. One woman explained that she moved to the rural area after the death of her husband because she could not afford to continue living in the city. Another respondent explained that he moved to the area soon after retirement because it was much cheaper to live in the rural areas. Those who cannot afford to live in the towns choose to settle in the rural areas because life is more peaceful and less complicated and has the added advantage of being less expensive. They are not heavily reliant on money for their survival in the rural areas as they do not have to pay for their accommodation and they can grow their own food. They are attracted to the rural areas because of its greater affordability. The high cost of living in the urban areas was an important reason for moving out of the town.

> After the death of my husband, I sold the few things we had in the house in town, then I came to buy this house here [in the rural area], because I could not afford the cost of living in town after my husband died, because he was everything to me....For example, life is very much cheaper here than it is in the urban areas. Houses are cheap and food we get from our fields and we do not have to pay for it like it is done in urban areas. We have access to free vegetables from the gardens, and we do not need to pay taxis to get to places like hospitals, market, or church, as we just walk.
>
> I love living here because it is where I have my friends and my land. I also live here because life is cheap, unlike in the town where I have to pay accommodation and buy food,

but here we do not do all those things. Although my house is very old and not up to standard, but at least I do not pay rent like in town.

….. I like it here because I have my land here and I do not have to buy everything like people who live in the towns. Yes I feel that I belong here because I have my family here, my land and my friends. Moreover, life in the city is very expensive so I love it here.

I just enjoy being here because it is where I own land and I was born here. I know nowhere else. For me, this is the best place although life is not easy here, but the fact that it is cheap compared to other areas. I like it. I like living in the village also because when you beg, people give you, but in the city when you are a beggar you die because no one gives you something for free.

It is clear from these extracts that many elderly choose to live in the rural areas. They have an emotional attachment to the land. They were born in the rural areas and they intend to die there. In addition, some have lived their whole life in the rural area and they cannot see themselves moving in their old age. Moreover, they have established social networks in the rural areas. Some also commented on the sense of community that exists in rural areas. The elderly are integrated into the social activities in the rural areas. They have formed deep and meaningful relations in the community and they are not willing to let them go. Others do not want to live in the towns because they have no social networks in the towns, especially those who were left alone following the genocide.

I belong here. I am happy with my son and my daughter in law; they love me and care about me. They are my source of life, and without them I can never survive here.

I love the rural area only because I am near my children, I can see them regularly, and I rely on them financially, so if I go far from them I can die because no one will look after me.

Some elderly are forced to live in the rural areas in order to ensure their independence. They do not have any form of support in the urban areas. In addition, they cannot afford the high cost of living in the urban areas. However, this does not mean that they are not aware of the advantages of urban living. Urban areas have better basic services and amenities than rural areas. In addition, there are more job opportunities and a greater variety of jobs which attracts the youth. Health facilities are often of a better quality in the towns and they tend to be closer to their place of residence. If they require specialised health services they have to travel to the town where there are doctors and hospitals that can cater for their health needs. This suggests that the population in rural areas feel more vulnerable because of the lack of basic amenities. Many blame the government for the failure to invest in rural areas. They express concern that rural areas are sidelined by the government.

I am bothered by three situations in my life. One is living alone in this house, I am always worried. The second is hunger because if my daughters do not come and bring me some food, I will be hungry for the rest of the days until they come, and lastly, is the concern about my health. I feel that our government has ignored older people in the rural areas and only considered those who live in the urban areas.

The thing I like about the rural area is that I own my own big house here, but what I dislike now is that people in the village are always forgotten by the government when it comes to service delivery, people in the main cities are the only ones considered. For example, we do not have electricity here, clean water, and tarred roads, but in the city these services have been delivered long ago, our children do not go to school because the nearest school is nearly 15 kilometres. So growing older in the rural areas means dying quickly because you do not get what you want.

High levels of poverty have forced many young people out of the rural areas. The unemployment rate in the rural areas is high because there are few economic opportunities. Young people migrate to the towns in search of better opportunities. They leave the elderly alone in the rural areas. Rural areas are often sparsely populated and as the young people leave this makes the elderly feel less secure. As more and more youth move to the urban areas in search of better opportunities, the elderly find themselves alone and fending for themselves.

> Poverty is the main factor which pushes them [youth] out of the village and they leave hoping for a good life in the urban area. Unfortunately, those who left the village before the genocide were all killed there during the genocide of 1994.
>
> They go to look for well paying jobs because there are no jobs in the rural areas, and some others go to further their education as there are no universities in most rural areas like here.
>
> The village is poor; they [young people] then go in search of jobs. The village remains almost uninhabited and less secure because of the absence of younger people. Poverty and unemployment are the major factors, which tends to affect us on their departure.

However, it is clear that the migration of the youth to the urban areas has had positive results for the elderly. Some of the elderly reported that their children had moved to the urban areas and they were sending back remittances to the household. These remittances were important to the survival of the elderly since many were not able to engage in physical labour, and as a result they could not grow their own food. It is however clear that for many of the elderly living in poverty, their children are an important source of support in old age. In Rwanda the rural economy is heavily dependent on agriculture which is labour intensive, using traditional farming methods. Older men and women were not physically strong enough to work on the land, therefore they were more susceptible to poverty. As people grow older they are less likely to be able to fend for themselves because of their weakened physical state. Only a few were in a position to carry out routine household chores or any other work including field cultivation. Years of poverty have also weakened them physically leading to their inability to function independently. As a result, they are heavily dependent on their children for their survival. However, not all received support from their children. Some reported that their children could not find employment in the urban areas. They were destitute in the urban areas and as a result they could not provide any support to their parents. Their parents were forced to survive on their own without any financial assistance from their children.

> When the youth leave the rural areas to go to the cities, the move is viewed negatively at first. They often go to look for well paying jobs that could maintain them. When they start sending home remittances, it becomes hard to see the negative influence of them leaving.
>
> They cannot resist the poverty. The village becomes insecure on their departure, but they have to go because some of them go to look for money and others just go there and become thugs, without any intention of returning home.

Older men and women noted that their physical health had deteriorated over the years and they were not physically strong enough to continue working on the land. Even carrying out household chores was physically challenging but they somehow managed with great difficulty. A few men and women stated that they were forced to cultivate the land because they were caring for children and/or grandchildren who were dependent on them for their survival. Some were no longer independent and were able to survive with the assistance of others. Some were dependent on their children for cultivating their land while others relied on their neighbours for assistance. The support they received from their children and others allowed them to grow their own food and ultimately ensured their survival. At the time of the interviews, older men and women stated that they were forced to be self-reliant because they did not have any children living with them. Their children had moved to the towns in search of employment. A few reported that their children had found employment in the towns but others remained without a job. Many reported that they had lost their children in the genocide. They expressed a sense of hopelessness for the future. They were particularly worried about surviving on their own because they did not have anyone to turn to especially in their time of need.

> I have a field and I still go to cultivate it, even when I am sick because if I do not, then these children will die with hunger.
>
> I cannot carry out most of the work especially when it is cold because I am asthmatic. The children carry out the work, although they cannot do it properly. Yes, I have a big plot of land but my children are the ones who cultivate it for me because I am sick. I am no longer able to cultivate the land.
>
> I am no longer able to do any work and here at home nobody helps me to do housework. I try to do as much as I can and that is how I live here in this house.
>
> I am no longer able to work and nobody helps me. I have land but it is cultivated by the neighbours because myself I am no longer able to do any work.

Among the numerous challenges respondents faced, they reported that the loss of close family members was the most devastating in their lives. All the older men and women reported that they had lost a close family member, most of the deaths occurring during the 1994 Rwandan genocide. Only few cases of deaths had occurred after the genocide, and these were due to natural causes or accidental injury. Elderly men and women reported that they were finding it difficult to cope with the death of their close family members. It was clear from their faces that the events of 1994 still continued to haunt them. As a result of their experiences they lived in a state of perpetual fear. They found it difficult to erase the memory of the events that killed people who were so close to them. For example, some elderly stated:

> I do not fully feel part of this community, because I do not have family living here in this community but I only have friends. All our children passed away during the 1994 genocide. This affected us deeply and we still live in fear.
>
> All my children left the village long time ago for the city, and many of them died there during the 1994 genocide and the rest I do not have their information because I never heard about them. I was left alone to deal with a number of challenges in my old age.
>
> This is my own house that I am now living in. It was left to me by my husband before he was killed in the 1994 genocide.

Almost all the men and women interviewed were affected by the bloody massacre in their country. They reported that it was difficult to erase memories of the genocide. They kept recalling the events that occurred during the genocide, and there was a sense of despair and helplessness among them. It was impossible to remain distant from those events because they not only lost their belongings but also close family members. Some reported that they had lost their spouses during the genocide while others had lost their children and grandchildren. Some had lost both their spouses and their children and they were now alone. Some also suffered injuries during the genocide that had long-term impacts on their health. One man reported that he lost his limb during the genocide and this negatively impacted his participation in society. Many of the men and women reported that they did not fully recover after the genocide and their health suffered. It is clear that they had lasting memories of the genocide because it led to the destruction of the social fabric of the community. It destroyed the sense of community that existed before the genocide. The genocide irrevocably changed their community because members were not as willing to help each other as they were before the massacre. The genocide left them with a feeling of distrust which continued to plague them and instilled a sense of fear in them.

> My health started deteriorating right after genocide because it is when my husband died in that war. At the time when my children were all still young, and it was hard for me to adapt to my new status and raise these children. This event affected my health negatively, and it is hard to control although now I am trying to forget and adapt to my new life status.
>
> ….before the genocide I was very happy with my wife and children. During the genocide, I lost my wife, and some of my children, and these changes have affected my health negatively. Since then I started having high blood pressure and other unknown sicknesses that I never had before.
>
> There are changes because before the genocide most of the sicknesses did not exist, but after the genocide which came with various forms of sexual violence, it affected many women badly and brought new infectious diseases that were not known before genocide. It had also created fear in us old people and took many of our younger people away from us, and this also affected our health negatively. For example now I am having heart problems. After the genocide I cannot stop thinking of my relatives that I lost in the conflict and I get sick.
>
> Before the genocide it was very nice. People loved and cared for each other, but these days it is not the case. For example, before the genocide, the nurses and doctors received the patients very well but nowadays, they first ask if the patient has health insurance, without which they will ask a lot of money before they can treat the patient. Thus nobody is willing to help another person anymore, our culture has changed drastically.

Of the total sample, 18 reported that they suffered from a number of ailments after the genocide that plagued their country. The in-depth interviews suggest that the elderly who survived the horrors of the genocide still suffer from psychological stress that increases their risk of health problems, such as high blood pressure. In the interviews, some also conceded that their health problems may have occurred as a direct result of the ageing process. Wagnild and Young (1993) observe that as people age, they often encounter challenges such as the development of chronic illness and the emotional stress resulting from the loss of loved ones. Some older men and women are also not very mobile. This means that they spend all their time at home without engaging in any physical activities, even walking. This does not mean that they are not able to engage in physical activities, but they choose not to because they feel they are too old.

Older men and women who experience ill health prefer to remain home rather than seek treatment at a health facility. For many, cost is one of the major barriers to obtaining health care. Older men and women reported that they do not visit health facilities because they cannot afford to pay for health care. The physical distance to the health facility is another barrier to accessing health care. The frail cannot walk a long distance to access health care services. The rural areas are characterised by hilly terrain which is also difficult to navigate. The long waiting period at health facilities is also a problem for frail elderly men and women. Given the challenges in accessing health care services elderly men and women rely on traditional healers to provide them with medicine. Those who live alone are particularly vulnerable because they are dependent on the assistance of others to get to a health facility. One example is of a male respondent who stated that when he was sick he was dependent on others for his survival.

> I go to the hospital or to any closest health centre when I am unwell. There are no barriers; we are very well received at the hospital when we get there, especially when you have money to pay. The only problem is that in order to see the doctors we have to wait in a queue and it is very long for someone who is ill to wait for so long and there are always so many patients.
>
> I take traditional medicine whenever I am sick, because I cannot afford medical doctors in hospitals and I do not like to add to the burden of my children who take care of me.
>
> Yes, I have barriers in wanting to access health services because I do not have health insurance, and this makes it hard for me. So I stay at home and wait for a 'Good Samaritan' to help me when I am sick.
>
> I go to bed when I am sick, and stay lying there until the day I will get well because I cannot afford money for medication. And the hospital is far, situated about 8 kilometres. I can hardly walk and there are no taxis here, and motorbikes are expensive.

10.6 Discussion

In the 1994 genocide in Rwanda, almost one million people were killed over the course of approximately 100 days. Although the exact death toll remains unknown, some estimate suggest that a large part of the country's entire population was murdered. In addition, estimates suggest that thousands of women and girls were brutally raped, while young children watched as their mothers and fathers were mercilessly murdered, often by people they had previously known as friends and neighbours. It has been more than 15 years, but the effects of the war are still felt in many ways by the elderly. Many of the elderly continue to live with the devastating effects of the war. They reported that they not only lost their belongings but also close family and friends. It is not surprising, therefore, that they are finding it difficult to move on with their lives. The death of loved ones has caused them a great deal of emotional distress. Many have lost their entire family and as a result they are living alone without any support. Loneliness is also a big problem for the elderly. It has also negatively impacted their health indirectly because they have no one to take them to the health centre if they need medical assistance.

The conflict not only impacted the health of the population but also the health system in Rwanda. In conflict situations, health facilities and systems are exposed to risks that could result in death and injury to health workers, damage or destruction of health facilities such as hospitals and community clinics, and disruption or elimination of health delivery systems (IRP and UNDP 2010). In Rwanda the genocide led not only to the destruction of the health infrastructure but also to medical staff leaving the country. With the advent of peace, there have been efforts to rebuild the health system. In 1995 the government issued a new policy to assist in the reconstruction of the health sector. In the rural areas, however, the number of health facilities remains woefully inadequate. Older people have to walk long distances in order to access health care, and this is not possible for the frail and physically weak elderly. Cost is also barrier to accessing health care. Poverty is rife among the population and the costs make health care services out of reach for many.

Some still live in fear for their lives and they continue to experience despair and hopelessness about their future. The study suggests that most of the elderly spend most of their days at home. They are not active for most of the day. They report that they are too physically weak to engage in domestic chores and other activities. Physical activity, considered labour by the elderly, has been identified as a protective barrier against a number of illnesses. Studies point out that physical activity has well-known benefits for several chronic disorders including coronary artery disease, stroke, diabetes mellitus, and osteoporosis. Moreover, its influence on premature mortality among both young and old segments of the elderly is well established, and there is existing evidence that the participation of the elderly in labour may delay cognitive loss and impairment (Hu et al. 2000; Ferrucci et al. 1999). The study found that the elderly often do not engage in any physical activities which may have exacerbated their poor health situation.

Rural areas are seen as more supportive of older people than urban areas. The community is seen as more supportive because people tend to know and care for each other. The rural elderly rely heavily on their social networks of family members, friends, and neighbours for assistance. In poorer communities, and especially in rural ones, the family is the primary source of support for older people, but in many cases its capacity to support older members is heavily limited. Some elderly men and women who were caring for young children or grandchildren reported that they were forced to work in order to ensure the survival of their household. This has resulted in a reversal of roles because the elderly now have to assume responsibility for the care of their children and grandchildren. In the past the elderly would have expected their children to provide for them. In Rwanda, the cultural concept of intergenerational support is reflected in the phrase, "urukwavu rukaze rwonka abana", literally translated, 'an old hare suckles from the young' (Marzi 1994, pp. 3–7). In the present study, the testimonies of elderly men and women suggest that they are struggling to provide care for the children in times of economic hardships.

Studies suggest that older people without support networks are particularly vulnerable to poverty (Barrientos et al. 2003). Rural areas are also seen as more affordable than urban areas. However, rural areas are also criticised for neglecting the health needs of the elderly. In many developing countries, the main focus of health care is on young children and their mothers, and seldom is there a focus on the health care needs of the older population. Studies in Africa suggest that the availability of, and geographical access to, health care services in rural areas, is severely limited (Barrientos et al. 2003). As people grow older and frail they become more heavily dependent on health care and the support of social services. Given that most rural areas are inhabited primarily by older people, there is a need for more effective coordination of major services in rural areas.

10.7 Conclusion

The literature suggests that globally, old age has been identified as one of the key causes of poverty. The situation of the elderly and the relationship between ageing and development remain important issues and deserve continuing and intensive attention and action (Andrews 1999). Governments in many developed countries have taken the initiative to introduce policies aimed at protecting the elderly from poverty. However, in most developing countries, the plight of the elderly often falls on the informal systems of care such as the extended family, but these are also under threat. This is one of the biggest challenges the elderly living in rural areas face, which makes them particularly vulnerable to poverty.

In the aftermath of the genocide, there have been some efforts to provide psychosocial support to the population. However, in resource-constrained countries that have undergone conflict, there are often limited resources to help build human capacity, promote social ecology, and strengthen the culture and values of a community, upon which psychosocial well-being is heavily dependent (Psychosocial Working Group 2004). Further, extreme poverty may also add to the feelings of hopelessness and lack of emotional well-being experienced by survivors of war and conflict. Almost two decades after the genocide there is still a need for health services that address the underlying trauma experienced by the elderly and its impact on their ability to function in society.

References

Andrews, G. R. (1999). Health and well-being of older persons in rural areas. Paper presented at the Fifth National Rural Health Conference, Adelaide, South Australia, 14–17 March.

Babbie, E., & Mouton, J. (2001). *The practice of social research*. Cape Town, South Africa: Oxford University Press.

Barrientos, A., Gorman, M., & Heslop, A. (2003). Old age poverty in developing countries: contributions and dependence in later life. *World Development, 31*(3), 555–570.

Bull, C. N., & Bane, D. S. (1993). Growing old in rural America: new approach is needed in rural health care. *Aging, 365*, 18–25.

Donovan, C., Bailey, L., Mpyisi E., & Weber, M. (2003). Prime-age adult morbidity and mortality in rural Rwanda: Which households are affected and what are their strategies for adjustment? Paper presented at the 25th International Conference of Agricultural Economists, Durban, South Africa, August.

Ferrucci, L., Izmirlian, G., Leveille, S., Phillips, C., Corti, M., Brock, D. M., & Guralnik, J. (1999). Smoking, physical activity, and active life expectancy. *American Journal of Epidemiology, 149*, 645–653.

Hampson, J. (1990). Marginalization and rural elderly: a Shona case study. *Journal of Social Development in Africa, 5*(2), 5–23.

Hu, F. B., Stampfer, M. J., Colditz, G. A., Ascherio, A., Rexrode, K. M., Willett, W. C., & Manson, J. E. (2000). Physical activity and risk of stroke in women. *JAMA, 283*, 2961–2967.

IRP and UNDP. (2010). Guidance note on recovery: Health. http://www.unisdr.org/files/18782_irphealth.pdf. Accessed 18 Oct 2011.

Krout, J. A. (1986). *The aged in rural America*. Westport, CT: Greenwood.

Krout, J. A. (1994). *Providing community-based services to the rural elderly*. Thousand Oaks, CA: Sage.
Madsen, W. (1999). *Genocide and covert operations in Africa, 1993–1999*. New York: Edwin Mellen Press.
Marzi, H. (1994). Old age in Rwanda: a problem? *Bold (Malta), 5*(1), 3–7.
Murthy, G. S., Gupta, S. K., Bachani, D., Jose, R., & John, N. (2005). *Current estimate of blindness in India*. India: BMO.
National Institute of Statistics of Rwanda (NISR), & Macro International, Inc. (2005). *Rwanda demographic and health survey 2005*. Calverton: Macro International.
Owen, W. F. (1984). Interpretative themes in relational communication. *Quarterly Journal of Speech, 70*, 274–287.
Penning, M. J., & Chappell, N. L. (1994). Health promotion and disadvantaged elderly. In C. N. Bull (Ed.), *Aging in rural America* (pp. 117–199). Newbury Park, CA: Sage.
Prunier, G. (1995). *The Rwanda crisis: History of a genocide*. New York: Columbia University Press.
Psychosocial Working Group. (2004). Considerations in Planning Psychosocial Programs. http://www.forcedmigration.org/psychosocial/papers/PWGpapers.htm. Accessed 18 Oct 2011.
Rwezaura, B. A. (1989). Changing community obligations to the elderly in contemporary Africa. *Journal of Social Development in Africa, 4*(1), 5–24.
Sangano, K., Nsanzabaganwa, S., & Mpyisi, E. (2003). *Regional workshop on ageing and poverty Dar es Salaam, Tanzania. Rwanda. Country Position Paper*. New York: United Nations Development Programme.
UNDP. (2010). *Human Development Report 2010. The real wealth of nations: Pathways to development*. New York: Palgrave Macmillan.
Wagnild, G. M., & Young, H. M. (1993). Development and psychometric evaluation of the Resiliency Scale. *Journal of Nursing Measurement, 1*(2), 165–178.

Chapter 11
Policy and Programme Responses

Ashley Gresh and Pranitha Maharaj

11.1 Introduction

> *"...population ageing can be seen as a success story for public health policies and for socioeconomic development, but it also challenges society to adapt, in order to maximise the health and functional capacity of older people as well as their social participation and security."*
>
> (WHO 2011a)

Population ageing extends beyond demographics having major social, economic, political, cultural, and public health implications, affecting nearly all countries in the world (McLigeyo 2002). As mentioned previously, the pace of population ageing is remarkably faster in developing countries than developed countries, which means that developing countries have less time to address the rapid pace of change and its consequences. With less time to adapt to the changes associated with population ageing, it is urgent that governments in Africa take steps to face the challenges and make the best of opportunities that population ageing can bring. Population age structures are shifting with an increasing older population, however not every person is reaching a secure, healthy, dignified old age (Sidorenko 2008). Unfortunately, issues associated with population ageing are often overlooked in Africa at the national level, and are under-researched. The steady growth of the older population is irreversible and is of special relevance in terms of public policy. This chapter explores the various responses of governments in Africa to population ageing. It looks more specifically at how the response has shifted, and the policies and programmes being developed and implemented to meet the health needs of older men and women. The majority of all older people currently live in low or middle-income

A. Gresh (✉) • P. Maharaj
School of Built Environment and Development Studies, University of KwaZulu-Natal, Durban, South Africa
e-mail: ashley.gresh@gmail.com; Maharajp7@ukzn.ac.za

P. Maharaj (ed.), *Aging and Health in Africa*, International Perspectives on Aging 4, DOI 10.1007/978-1-4419-8357-2_11, © Springer Science+Business Media New York 2013

countries, highlighting the need for new thinking and new policy responses to population ageing in developing countries.

The ageing of populations in Africa is rapidly progressing against a background of developmental, infrastructural problems such as widespread poverty, low levels of education, lack of social security, violence and armed conflict, as well as inadequate health and social services (Bongaarts 2005; HelpAge International 2003; Joubert and Bradshaw 2006; Kalache 2008). As the demographic transition is occurring in most African countries before the introduction of any significant socioeconomic changes have been made, policy development, implementation, and resource allocation is particularly difficult (Joubert and Bradshaw 2006). Many studies indicate that older people's lives are now marked by inadequate family support due to changing social and economic circumstances, chronic poverty, ill health, and isolation from health services (Aboderin 2006). The older population is particularly vulnerable to these major developmental issues facing the African continent, in addition to the expected physical, mental, and physiological changes associated with the ageing process (HelpAge International 2003).

The majority of public health programmes across the African continent face a "triple burden": resource constraints, prevalence of infectious and rising chronic disease, and conflict, injury, and trauma (Aboderin 2010). Africa throughout history has carried a high burden of disease such as malaria and tuberculosis. Currently it is the epicentre of the AIDS pandemic, and is still plagued by a high prevalence of infectious diseases (Cohen and Menken 2006). At the same time the prevalence of non-communicable diseases (NCDs), cardiovascular diseases, cancers, diabetes, and chronic respiratory diseases often associated with ageing, is increasing (CNCD 2011). By 2020 it is projected that three quarters of all deaths in developing countries will be age-related (McLigeyo 2002). As African countries transition from focusing on infectious diseases to a dual focus on communicable as well as NCDs and an accelerated rate of population ageing, it is necessary to effectively develop and implement policies to prevent and control NCDs (Giles 2010). The cost of health care is expected to rise substantially due to the double burden of diseases in the ageing population. Although the AIDS pandemic is projected to reduce life expectancy in many African countries, the older population is still continuing to grow and is expected to grow and effectively triple in size from 5% of the population to 15% in 2050 (Bongaarts 2005). Often the focus is on a "crisis" of population ageing rather than a challenge for policies to be developed that avert the negative consequences associated with this major demographic change (Randel et al. 1999). There is a need to shift thinking from the former ideas that: population ageing is mainly an issue in developed countries, older people represent an unproductive burden on society and a hindrance to social and economic development, population ageing will put unsustainable pressures on formal social protection mechanisms, and that the health of older populations is not a priority issue for public action (Lloyd-Sherlock 2002; Dullemen 2006).

Older people play a vital role in African society; however, they are often excluded from development programmes for the above reasons and discriminated against by health care systems (HelpAge International 2008b). Countries can benefit from what is referred to as a "demographic dividend", a large increase of economically active adults who enter the workforce as fertility declines. If investments in education, health, and economic opportunities for adults are expanded, this rapidly increasing number of adults have the potential to provide a catalyst for national economic growth and development (Bremner et al. 2010). Studies from countries such as Ghana, South Africa, and Brazil have shown the remarkable social and economic contributions older people make to their families and communities (Dullemen 2006). This is especially apparent in the context of HIV/AIDS, where it is documented that older persons become the primary caregivers of afflicted children and the orphaned grandchildren left behind by deceased children (WHO 2002). The well-being and health of older people is inseparable from wider processes of development, and there is a growing call for a shift in thinking to focus on the actual and potential contributions of elders based on their human rights.

There is a general fear that the ageing population will create an unmanageable increase in demand on health care services and social security costs. However, there is evidence to suggest that cooperation from multiple sectors, advance planning, innovation, and making evidence-based, context appropriate policy choices will enable countries to successfully manage the basic health needs of the ageing population (WHO 2002). Other evidence shows that the impact of population ageing on health spending can be mitigated by the overall organisation of the health care system, and the solutions lie in the broader context of health care financing and organisation (Lloyd-Sherlock 2000). Health is a priority concern in order to address issues of population ageing because good health is directly connected to the ability to work and sustain a reasonable standard of living for oneself and one's dependents, and is the key to creating and maintaining a productive society to fuel national development (Randel et al. 1999).

11.2 International Responses to Population Ageing: A Focus on Health Care

Population ageing as a global issue has a relatively short history. The issue of ageing was first raised in 1948 when Argentina submitted a draft declaration on old age rights to the United Nations General Assembly. The draft was not adopted but was referred to the Economic and Social Council. Then more than two decades later in 1969 the issue was once again placed on the agenda on the United Nations at the initiative of the Government of Malta culminating ultimately in the first-ever World Assembly on Ageing in Vienna, Austria. This was accompanied in the same year by the adoption of the International Plan of Action on Ageing which has served as the international blueprint for the development of policies and programmes on ageing. It included recommendations for governments on health and nutrition, family, and social welfare. At that time it was apparent that populations in developed countries were ageing but this was not a major concern for developing countries. In 1990 the United Nations General Assembly declared 1 October as the International Day of

Older Person, an event now celebrated annually in a number of countries. In 1991 the United Nations Principles for Older Persons was formally adopted by the General Assembly, which was seen as the second-most significant development in international policy action on ageing. It emphasises dignity, self-fulfilment, participation, independence, and care of older people and encourages national governments to include them in programmes (United Nations 1991).

In the 1990s a new approach to ageing was starting to emerge, strongly influenced by the World Health Organization (WHO). In 1995, WHO changed the name of its "Health of the Elderly Program" to "Ageing and Health" marking an important change in its orientation and shifting the response to issues of population ageing and health (Kalache and Sen 1999). The new name emphasises a life course perspective and an understanding that the health of older persons can only be understood in the context in which they live, as well as their life experiences. The aim of the programme was to develop policies that ensure: "the attainment of the best possible quality of life for as long as possible, for the largest possible number of people" (Kalache and Sen 1999). This marked an international shift in thinking about population ageing issues in conjunction with health care, highlighting the importance of older persons as a population group.

In 2002 WHO developed the concept of "active ageing" which is defined as, "the process of optimising opportunities for health, participation and security in order to enhance quality of life as people age (for both individuals and population groups)" (WHO 2002). According to WHO, health refers to "a state of complete physical, mental and social well-being, and not merely the absence of disease." The framework emphasises the need to realise physical, social, and mental well-being throughout the life course. "Active" refers to not only the ability to be physically active, but also the ability to contribute to national development, communities, families, and peers. A key goal is maintaining autonomy and independence for the older people. Active ageing is a rights-based approach to ageing issues based on the recognition of fulfilling older people's rights following the United Nation's Principles for Older Persons (United Nations 1991).

The "Active Ageing Framework" outlines suggested policy responses based on three pillars of action. The first pillar is health. WHO first addresses the need to prevent and reduce the burden of disabilities, chronic diseases, and premature mortality, which all involve addressing:

- Economic influences on older persons' health (poverty, income inequities, and social exclusion, lack of education)
- Age-friendly, safe, clean environments (in and outside of health facilities)
- Enhancing older persons' quality of life
- Social support through community outreach
- Impact of AIDS on older people
- Mental health

Then it focuses on the need to reduce risk factors associated with major diseases, and promote health throughout the life course looking at tobacco use, alcohol and drugs, physical activity, oral health, nutrition, and access to essential safe medications.

It also emphasises developing a continuum of comprehensive health and social services to meet the needs of men and women as they age. In addition, countries should provide training and education on gerontology to both formal and informal caregivers.

The second pillar of the "Active Ageing Framework" is participation in society to ensure that older people are involved in the labour market, employment, education, health and social policies and programmes. The older population plays a major social and economic role in society and also constitutes a growing political constituency. Recognising and harnessing their full participation will benefit the health and well-being of individuals, families, communities, and societies. The third pillar is security, to ensure that the social, financial, and physical security needs and rights of people as they age. Failure to ensure the security of older people jeopardises their health and their productive contribution to society. All of these pillars are connected to meet the needs of older men and women.

In 2002 the Madrid International Plan of Action on Ageing (MIPAA) marked a turning point on how the world addresses the challenges of a rapidly ageing population (United Nations 2009). The MIPAA evolved from the 1982 World Assembly on Ageing in Vienna, and the 1991 United Nations Principles for Older Persons mentioned previously. The MIPAA calls for changes in attitudes, policies, and practices at all levels and across multiple sectors so that population ageing issues will be addressed (United Nations 2002). The plan made a political commitment to incorporate ageing within social and economic strategies, policies, and actions, based on the recognised need for a shift in policy in response to increasing numbers of older persons globally.

The conceptual framework that guided the MIPAA was that of a "society for all ages". This was the first time that governments adopted a comprehensive approach linking issues of ageing to socio-economic development and human rights (United Nations 2009).

The MIPAA outlined three priority directions for action:

1. *Older persons and development*: to integrate ageing within the context of development to ensure that older persons are active participants in society and maintain peace, health, and security
2. *Advancing health and well-being into old age*: recognising that good health is vital to the development of individuals and the overall health of the population, and vital for economic growth and national development; there is a need to ensure access for older persons to preventive and curative care and services involving health promotion and disease prevention
3. *Ensuring enabling and supportive environments*: to promote positive perceptions of ageing in order to effectively implement all programmes and policies, and foster lifelong education, skills upgrading, and healthy lifestyles (United Nations 2002)

These directions for actions have the aim the development of ageing populations and simultaneous socio-economic development with an involvement from multiple sectors. The MIPAA has a strong development focus, arguing for the mainstreaming of the process of global ageing within the larger process of economic and social development (United Nations 2002). The MIPAA came to serve as the basis for future plans of action on ageing around the world.

In 2003 a similar action plan was established more specifically for the African continent: the African Union Policy Framework and Plan of Action on Ageing (AU-Plan). This framework was based on the previous international frameworks outlined above. The AU-Plan calls for the involvement of all stakeholders at all levels including individuals, communities, NGOs, donors, the private sector, the media, governments, and other civil society groups. The goal of the AU-Plan is to guide all member states as they design, implement, monitor, and evaluate integrated national policies to meet the collective and individual needs of older people (HelpAge International 2003). The AU-Plan recommends that all member states address the following issues:

1. *Rights*: recognise the fundamental human rights of older persons and eliminate age-based discrimination and enable them to access all of their rights
2. *Information and co-ordination*: research and collect information about the situation of older people, as the gap in data inhibits effective allocation of resources to meet the needs of the older population; ensure the needs and rights of older people are integrated into all existing and new policies in all sectors so that older people are involved at all levels of policy development, implementation and monitoring and evaluation; establish coordinating and monitoring mechanisms to ensure issues of older people are addressed
3. *Poverty*: promote poverty-reduction strategies as older people are heavily affected by structural adjustment programmes and are most often the hardest hit by cuts in social welfare programmes
4. *Health*: ensure access to health services; meet the specific needs of older people, manage chronic diseases; provide free health services for older people; train health personnel in gerontology; provide education on healthy lifestyles
5. *Food and nutrition*: most programmes of food and nutrition focuses on the needs of under fives, mothers and younger populations, but there is a need to assess the nutritional situation of older people in order to ensure access to the means of food production and marketing
6. *Housing and living environments*: access to safe, durable, and affordable shelter
7. *Family*: promote and strengthen the role of the family and the community in the care of older members, and empower older members to contribute to their families
8. *Social welfare*: while social security legislation exists, it does not always give consideration to older people and is often centralised so older people cannot access it
9. *Employment and income security*: establish formal and informal social security systems
10. *Crises, emergencies, epidemics*: ensure assistance reaches older people in situations of conflict and crises; protect rights and needs of older people affected by AIDS and other epidemics

11. *Ageing and migration*: respect migrant worker's rights to employment
12. *Education and training*: ensure access to education and training for older persons; encourage their participation as educators and trainers
13. *Gender*: mainstream ageing into policies relating to gender (HelpAge International 2003)

This was the first time that the AU explicitly recognised the need to mainstream ageing issues into national policies. In terms of health, the AU-Plan of 2003 recognises that older people's capacity to earn a living and participate in community life is dependent to a large extent on their health status. In addition it outlines the situation of older persons in Africa documenting that older people are often denied access to essential health services, and few specialist services exist for them.

It also acknowledges the positive role that older people can and do play as providers of traditional medicine and caregivers of family and community members. In many African countries health care is also provided outside of health facilities by traditional healers. In sub-Saharan Africa traditional healers outnumber allopathic medical practitioners by more than 50 to 1 (Cohen and Menken 2006). Often traditional healers are older persons and acknowledged by communities as health care providers administering home remedies, and they are proportionally higher users than other population groups (Kalache and Sen 1999; HelpAge International 2003). The AU-Plan recommends developing strategies for the provision of safe traditional medicine.

The AU-Plan also highlighted another important aspect of ageing: a focus on gender. Research needs to reflect the different health issues affecting older women and men. Increasingly attention is being paid to the fact that women make up a greater proportion of older persons and health and welfare needs should be specifically targeted for them. The majority of older persons in African countries are women, with the women to men ratio increasing with age (United Nations 2009). Women are disproportionately affected by high levels of poverty and economic dependence, are victims of war and violence as well as experience inequities in nutrition and food distribution, education and have limited decision-making power (Kalache and Sen 1999; Ewing 1999; McLigeyo 2002).

The health of older women is often neglected or ignored, and it is usually older women that are caregivers of households, particularly in the context of HIV/AIDS, taking care of children and orphaned grandchildren (McLigeyo 2002; Aboderin 2010). Ageing affects women and men differently, not only physiologically but also socially (HelpAge International 2003). For example, physiologically women are affected by conditions such as post-menopausal reproductive health problems, and other reproductive health issues that have negative impacts on them later in life. Socially women are often denied property rights, which, for many widows, mean the loss of property and home. Therefore it is important that gender should be integrated in all international and national policies and programmes related to older persons.

Aside from the AU-Plan, international attention has been given to older women at the World Conference on Women in Nairobi (1985) and Beijing (1995) as well as the 1994 International Conference on Population and Development (ICPD) held in Cairo. Recognising the value of women's contributions to families, communities, and society, it is crucial for societies to make efforts to improve women's health and well-being in order to achieve sustainable development (Aboderin 2006). The agendas in Nairobi, Beijing, and Cairo highlighted older women as a particularly vulnerable group. The Nairobi agenda called for policies to provide social insurance for older women, and the Beijing agenda targeted women in anti-poverty programmes and to ensure health services focused on chronic conditions affecting women. In addition, the Programme of Action of ICPD called for governments to seek to enhance the self-reliance of the elderly population in order to facilitate their continued participation in society, paying special attention to the needs of elderly women. Gender should be focused on throughout the life course.

Both the MIPAA and AU-Plan call for multi-sectoral health promotion strategies to prevent disease and disability among older persons as well as to ensure access to adequate health services, both curative and rehabilitative, with the ultimate goal of enhancing older persons' quality of life. By signing these frameworks, African governments are acknowledging the need for policy responses for old-age health needs (Aboderin 2010). Health can only be sustained through the participation of multiple sectors (WHO 2002).

Other international agreements that are non-age specific should also be upheld to support the rights of older persons, such as the United Nations Declaration of Human Rights (1948), African Charter of Human and People's Rights, International Covenant on Economic, Social and Cultural Rights (ICESCR), and the UN Standard Rules on Equalization of Opportunities for Persons with Disabilities (1996).

In addition to these international policy calls, international nongovernmental organisations (NGOs) and civil society also have a large role to play in advocacy and assisting the implementation of programmes and policies. One example is HelpAge International. Since its inception it has taken an active role in advocacy for the older population around the world. It also works closely with governments to develop and implement policies and programmes to address the needs and rights of older persons.

11.3 National Policy Responses

Despite these international policy calls and expressed commitment on the part of many African governments, comprehensive national policy action is lacking (Aboderin 2010). In 2007 an appraisal of MIPAA was conducted among select African countries and the findings show that few specific strategies have been developed, ratified, or implemented across signatory countries. Policies targeted at old age or mainstreaming issues related to old age are at differing stages of implementation across countries in Africa.

In six African countries, there are national policies on ageing. In 2003, Tanzania formally adopted a National Policy on Ageing. In 2004, Mozambique also adopted a National Policy on Older Persons. In 2006, South Africa's Older Persons' Policy was approved by cabinet. In 2009, the National Policy for Older Persons was adopted in Uganda and Kenya. Most recently, in Ghana the National Ageing Policy was approved in 2010 (UNFPA and HelpAge International 2011). In other African countries (such as Cameroon, Malawi, Nigeria, Rwanda), there are preparations to introduce a policy on ageing or they have already been drafted but not yet formalised (UNFPA and HelpAge International 2011). A couple of countries have drawn policy guidelines aimed at initiating and formulating national plans of action. For example, the Egypt Strategy and Action Plan on Ageing 2007, which makes direct mention of the Madrid Plan, provides for the development of a strategy for control of NCDs, particularly for the older population (UNFPA and HelpAge International 2011).

Improved health systems have the potential to meet older persons' individual health needs as well as improving overall socio-economic development by increasing the productivity of entire households, as well as improving the efficacy and efficiency of interventions targeted at older persons. However, health systems are still largely inaccessible to older persons (Aboderin 2008). There is a need to shift policy responses toward addressing problems of NCDs such as hypertension and other conditions associated with the rapidly growing ageing population. A re-evaluation of health services should consider new approaches to chronic care, health prevention and promotion measures, and capacity building, with an emphasis on primary health care to meet the needs of population ageing.

11.4 Impediments to Policy Development and Implementation

The global economy and its effects on individual countries should be taken into account when considering policy development and implementation with regard to population ageing. The recent global economic recession's (2007) unfolding has highlighted the vulnerability of older populations with the loss of retirement savings and pensions, as well as causing a decrease in public expenditures such as health care (Bremner et al. 2010). The financial crisis reduced the value of pension funds around the world, causing a large negative rate of return on pension funds' investments, and a reduction of assets for older persons as well as reduced employment (United Nations 2009). This has major implications for allocating resources to implement policies on ageing, as it has in many countries shifted a focus away from older persons' issues.

The majority of African countries are particularly vulnerable to the financial crisis as they are economies in transition, and often have a weak infrastructural base. Most African countries are still largely dependent on subsistence agriculture, and the average gross national product (GNP) income per capita is currently lower than it was at the end of the 1960s (Cohen and Menken 2006). Due to the systemic nature of the current financial crisis, families are unable to serve as a buffer from the negative shocks in the economy for older persons due to the overall reduction in income. This makes a case for increased attention to older persons to counter the impacts of the economic crisis. It also creates a challenging environment for enacting policies from a global context trickling down to a national context. Africa also contains a growing share of the world's poor.

There is a lack of basic agreement on the definition of who an older person is, which is crucial to create policies, and compare and study countries across Africa. While developed countries have defined "elderly person" as 65 years old, this does not fit well for the situation in Africa (WHO 2011b). There is no standard definition of the older population, and most default definitions refer to the age when one becomes eligible for occupational retirement pensions, however, this does not always work in contexts where people do not work in the formal sector, there is high unemployment and low life expectancy. Most African countries define the "elderly" between 50 and 65 years of age depending on the country and the region (WHO 2011b). WHO defines older persons as 60 years of age and above, which is used in many African countries and will also be used throughout this chapter.

While there is no consensus about the definition of old age, a significant issue in many African countries is that many older persons do not possess identification cards (IDs). This is another major barrier to accessing health care and implementing policies related to old age. In Mozambique, a survey found that 42% of older people have no ID cards which are needed in order to claim free health care (HelpAge International 2008a). Tanzania faces similar problems of older people proving through ID cards that they are older than 60 years of age to receive free health care (Ministry of Labour, Youth Development and Sports 2003). The requirement that an applicant must be in possession of an identity document in order to access services may cause considerable hardship. Some older people may find the bureaucratic process of obtaining the required levels of documentation difficult. In a study conducted in South Africa, respondents described their experiences in trying to obtain identity documents as time consuming, expensive, and protracted (Padarath et al. 2006).

In addition to a lack of consensus on the definition, there is only a limited understanding of the magnitude, patterns, social determinants, and impacts of ill health on older ages across Africa. Some refer to this information gap as a "time bomb", as population ageing is occurring at a fast pace, irreversibly; therefore, governments must be sensitised of the consequences and implications of this phenomenon (Mba 2010). The average sub-Saharan African country spends 5.5% of Gross Domestic Product (GDP) on health care, of which most is spent on tertiary care (Cohen and Menken 2006). To improve health systems and the allocation of resources, it is crucial for policy makers to understand the nature of health problems that older age groups face as well as the changing disease patterns. There is a giant gap in understanding the changing health profile of older people and its implication on the demand for health services.

The determinants of health, such as individual, biological, and psychological factors, physical and social environments, and economic determinants need to be well understood to create context-appropriate policies that are effective for each individual country. In order to enact policies related to known implications of ageing, such as the increasing prevalence of disability and NCDs, there is a need to understand the risk factors and prevalence, as well as the socio-economic costs. There is also a need to gather data on the level of basic service provision given to older persons, and the share of the national health services that older people use, their access to health care and health service utilisation, and how these issues impact countries' social and economic development. The existing information gap on population ageing and its implications for health systems, compounded with the lack of definitions, creates barriers for policy development and implementation. Some hypothesise that if information on ageing and health is available, this has the potential to serve as a catalyst for national governments to generate the political will to act (Aboderin 2010).

However, some studies have been done that outline the health profiles of older people in sub-Saharan Africa, such as the WHO Study on Global Ageing and Adult Health (SAGE) study, which conducted studies in Ghana, South Africa, India, China, Mexico, and Russia. The study's findings show that health significantly impacts older people's quality of life, and that older persons systematically have less access to health care and lack access to the required curative, preventive, or diagnostic health services (Suzman 2010). The findings state that ill health varies between socio-demographic groups, such as gender, rural versus urban, and socio-economic status, and the prevalence of cardiovascular and NCD risk factors vary across countries as well as socio-demographic groups (Aboderin 2010).

The lack of political awareness and will surrounding the need for action on ageing underlies many countries' reasons for lack of policy development and failure to ratify and implement policies for ageing. Older persons' issues are a lower priority compared to others, such as economic development linked to major industries (Asagba 2005). The lack of political awareness also reflects a lack of pressure from civil society and advocacy efforts from practitioners. For example, in Nigeria, the Ministry of Health formulated a broad National Policy on the Care and Well Being of the Elderly (finalised in 2003), which follows the guidelines outlined in the MIPAA, however, no government funding was made available for the policy (Asagba 2005). Nigeria's policy outlined the need to relieve older people of poverty, ill health, and malnutrition, as well as the impact of HIV/AIDS. Not only has this yet to be ratified, but the government has also obstructed the enactment of the Revised National Health Policy (2004), as well as the National Policy on Ageing presented again in 2008 (Aboderin 2010).

The above-mentioned information gap again underlies the lack of political action taken, as policy makers do not have the information as to what specific measures are needed to ensure the health of older adults. A lack of insight into how health and other sectors can appropriately respond to older persons' health priority needs hinders policy development, and leads to incoherent policies. For example, in Kenya, the National Health Sector Strategic Plan II 2005–2010 lacked specificity in terms of its efforts to incorporate older persons in its new policies and programmes (Aboderin 2010). In Zimbabwe, only fragmented and incoherent pieces of legislation exist for the welfare of older people; there is no comprehensive social protection scheme, making it difficult for older people to survive (UNFPA 2002). In Botswana, a study found that policies are not consistent; a review of the health system found that 83% of elderly hypertension patients interviewed indicated they paid nothing for services received, while 16% said they paid for services, indicating that there is a need for social security policy that clearly indicates the health benefits for the elderly (Botswana Ministry of Health n.d.).

Economic and political factors heavily influence the development and implementation of health policies for the older population. In addition, it should be noted that violence and conflict are also major hindrances to the development of nations. In Angola, for example, the health situation has deteriorated since 1980 because of the socio-economic crisis in the country, as well as the intensification of the war, which has restricted the budget allocated to the health sector (Castelo et al. 1999). In Sudan, some studies show that in natural disasters, political and civil conflicts, older persons are the last to receive assistance (Bashir et al. 2007). The Sudanese government has yet to address older persons' health needs. Some countries, like the DRC, Burundi, Sierra Leone, and Sudan are emerging from decades of mismanagement, political strife, and instability, but still more efforts are needed to further promote peace and reconciliation as well as improve governance.

In general, there are two types of policies that are needed: age-specific policies to address the individual needs of older persons, to ensure health, independence, participation, self-fulfilment and dignity, and age-mainstreaming policies that integrate collective issues of ageing into national development strategies (Sidorenko 2008). In order to ensure that policies are effective, nations need sound institutional infrastructure with collaboration from all major stakeholders including both the government and civil society. Human resource development should occur to train professionals to deal with ageing issues, specifically formal and informal health care workers with skills and knowledge of ageing and the needs of older people. And finally, financial resources are needed to allocate to programmes and social security systems to support older persons.

11.5 Age-Specific Policies and Programmes for the Health of Older Persons

Some African countries have created policies that allow individual older persons access to health services following the international calls for policy action for the improvement of older persons health. The following sections give examples of such policies from countries such as Ghana, Senegal, South Africa, and Tanzania.

In Ghana, under the National Health Insurance Scheme (NHIS), there is an exemption policy for those over 70 years old which benefits older persons with: in-patient care, emergency and transfer services, as well as out-patient care and coverage of the top ten national diseases (NPC 2007). With the introduction of the NHIS, older persons are able to benefit from free health care without making contributions to the scheme. However, studies indicate that there are problems of awareness of older persons' rights, with a survey that found that most older persons were unaware that they were exempt from paying user fees for services in public hospitals, which greatly reduced accessing care (Lloyd-Sherlock 2000). Once implemented, it is important to ensure that older people are aware of their rights and benefits in order to utilise these free health care services to improve their health.

In Senegal health care is not free for the entire population. In 2006 "Plan Sesame" was introduced to give all people age 60 years and older free health care. The Plan was created recognising the importance of older persons' health and rights, and to emphasise intergenerational solidarity. "Plan Sesame" was developed as a collaboration between the IPRES (Institute of Retirement of Senegal), the FNR (National Retirement Fund), the faculty of medicine of Dakar, and local collectives in order to enact health services for the older population (Coumé 2009). However, as health conditions improve throughout the nation and there is an increase in life expectancies, some concerns are that sustaining the plan and providing long-term financing for it will be a challenge due to an increased demand on health services from the old-age population (Oxford Business Group 2008).

South Africa also provides free health care for all older persons, as well as free health care for beneficiaries of the old-age grant. Older persons have free access to primary health care throughout South Africa, and recipients of social grants have, in addition to free primary health care, free access to secondary health care services at public hospitals (Joubert and Bradshaw 2006). There are three specialised geriatric departments that exist in the country. While in principle these services are provided for free to older persons, one qualitative study revealed that older people report a shortage or unavailability of essential medications, an inability to access assistive devices, and a perceived lack of respect, thoroughness, and information sharing from health professionals that attend to them (Joubert and Bradshaw 2006).

Tanzania provides free health care for persons older than 60 years of age. However, the majority of older people residing in rural areas are unable to access health services because of their inability to prove that they are 60 years and over. The National Ageing Policy (2003) addresses this problem by adjusting the criteria for determining that people are 60 years and above to allow access to health facilities. In addition the National Ageing Plan highlights the need to train health professionals for older persons' care in order to provide comprehensive care, targeting the health needs of the ageing population (Ministry of Labour, Youth Development and Sports 2003).

Mauritius, in addition to free health care, has established services outside of health facilities for the social, recreational, and health needs of older persons. Health and nutrition clubs have been set up throughout the country to promote healthy lifestyles for the elderly and their families (Sunkur 2007). There are also specialised geriatric sessions at health clinics for older persons with separate lines for them to collect their medicine at hospitals. Tunisia also has a social security system that is linked to health care provision; the social protection policies include national health and education provision (HelpAge International 2008c).

These policies indicate that, while old-age-specific policies offer older persons free access to health care, it is not enough to ensure that older persons are utilising these services. There are a number of barriers to accessing free health care services. This brings to attention the need to mainstream ageing issues into broader policies and programmes to ensure that health care is available, accessible, and affordable for older age persons.

11.6 Age-Mainstreaming Policies and Programmes for the Health of Older Persons

Many countries regard primary health care provision as the priority for health care delivery for the entire population, including all vulnerable groups, such as older age persons. The shift in thinking now advocated by international action plans calls for an integration of old age issues into overall economic, social, and political development strategies. The following section will outline various policies and programmes that are integrating old-age health issues into larger development plans in such countries as Ghana, Kenya, Mauritius, and Nigeria.

Kenya has subsidised primary health care for all citizens, but there is no particular focus on older people, nor is there information on the health profile or needs of older persons. However, Kenya's NHSPP II does make mention of the older population (60 years and older) saying:

> Various measures will be taken to improve financial access to health services, specifically for the financially vulnerable and the very poor: the elderly, street children and orphans, single mothers, and patients with chronic diseases like TB, HIV/AIDS, diabetes, etc. (Ministry of Health 2005).

The NHSPP II also outlines the need for annual screening and medical examinations, promotion of general hygiene, and social/emotional/community support for the elderly population.

Nigeria also has subsidised primary health care for citizens, but again there is no specific focus on older persons as of yet, and there is no state provision of care for the elderly. Mauritius is a welfare state and health care is free for Mauritian citizens. In 2001, the Ministry of Social Security, National Solidarity and Reform Institutions introduced the National Policy on Elderly which has the theme "Ageing with dignity". Mauritius is one of the first countries in Africa to devise such an official Plan of Action for the elderly. A well-organised network has been established in Mauritius to ensure cost-effective care for older persons involving partnerships between government and NGOs (Sunkur 2007). A geriatric centre was established in Senegal for the region of West Africa. It was created to act as a reference centre for geriatric issues and to employ the use of telemedicine to communicate to other parts of Western Africa (Coumé 2009).

In Ghana, in addition to the exemption for free health care for older persons, other ministries have also taken action to include older population groups in their development strategies. For example, the Ministry of Women's and Children's Affairs initiated activities such as *Gambaga* to support women with money, food, clothing, and food processing machines as well as micro-credit schemes. These initiatives set out to address the problems of old-age social isolation and exclusion, as well as the gender aspect of ageing. *Gambaga* is a village that has traditionally been a place where women, the majority elderly, are in exile from their families and communities after being accused of witchcraft. In Burkina Faso, older women also face accusations of witchcraft, leaving them excluded and often exposed to violence and abuse (HelpAge International 2010).

There are various programmes and interventions that are surfacing from different sectors that are beginning to take into account older persons, integrating them into development agendas. Civil society initiatives are also prevalent and will be discussed in this chapter. One of the main forms of mainstreaming ageing policies into development agendas is in the form of social protection measures, which will be addressed in the following section.

11.7 Social Protection for Older Age Groups and Its Implications for Health

> "The biggest enemy of health in the developing world is poverty." Kofi Annan (WHO 2011c)

Health, old age, and poverty are inextricably linked. Economic barriers for older people to access health care include: the inability to pay user fees, the inability to pay for transport, the fact that many are physically unable to reach a health service due to inability to purchase assistive devices, and the lack of ID cards (HelpAge International 2008a). In addition, poverty fosters ill health, forcing people to live in unhealthy environments, without shelter, clean water or adequate sanitation (WHO 2011c). Poverty also exacerbates the effects of ageing such as hypertension, malnutrition, anaemia, and other NCDs. Addressing issues of poverty in association with health care is crucial to ensure that older people as well as those they support are able to access health services and maintain healthy lifestyles.

The concept of social protection is receiving increased attention as a strategy to address issues related to poverty and older persons' vulnerability (Cohen and Menken 2006). In 2006, the African Union Commission, discussing social protection schemes across Africa in Livingstone, defined social protection:

> Social protection encompasses a range of public actions carried out by the state and others that address risk, vulnerability, discrimination and chronic poverty. The right to social security in childhood, old age and at times of disability is expressed in a range of international human rights declarations and treaties. Social security transfers in the form of, for example, pensions, child benefits and disability allowances are considered to be core elements of a comprehensive social protection system (HelpAge International 2008c).

Currently, according to the World Bank, globally only one quarter of the labour force is accruing pension rights, and four out of five older persons do not have any pension coverage (United Nations 2009). In most African countries one out of ten older people receive a pension (PensionWatch 2011). One in five of the poorest in the world, living on less than a dollar a day, are over 60 years of age (Dullemen 2006). A survey of 15 African countries found that in 11 countries the proportion of older people living in poverty was higher than the national average, making them a particularly vulnerable population (HelpAge International 2008b).

As seen earlier, international calls for action suggest that older people are in need of a basic income either generated through employment or social cash transfers such as social pensions, disability grants, or childcare grants (HelpAge International 2008a). Social cash transfers have the potential to not only reduce old-age poverty, but also intergenerational poverty; if the money is spent on the household it can lead to improvements in education for children and future cohorts of older persons (PensionWatch 2011). Some studies show that pensions and grants for older people give them the ability to invest in the education of their children, buy food, and access and pay for health services (HelpAge International 2008a). Social pensions (non-contributory cash transfers paid regularly to old people) have the potential to be one of the most effective tools to invest in human capital development and reduce poverty (PensionWatch 2011). Evidence from long-standing social protection systems show that social protection improves health, nutrition, and social status across the life course, and ensures regular support for households affected by AIDS (HelpAge International 2008c). A study in South Africa found that, in households that pool income, the old-age pension protects the health of all household members, working in part to secure the nutritional status of household members, in part to improve living conditions, and in part to reduce the stress under which the adult household members negotiate daily life (Case 2001). Improved health and nutrition can further improve economic productivity in the short term and can eventually help them break the cycle of poverty in the long term.

However, there is much discussion on what social protection measures should be put in place. The debate is shifting from not if there should be social protection, but how (HelpAge International 2008c). A central question around the debate of social protection measures is whether or not countries should establish a universally applicable pension system for retired older people (UNFPA 2002). Some view the universal pension system as an adoption of a Western style system that may not be suitable for developing societies. Formal social security schemes in Western countries are able to redistribute wealth and reduce poverty and inequalities between sectors of society. However, in many African countries social security systems are not as effective, due to the small percentage of the labour force that is covered and the size of the pensions received (Cohen and Menken 2006). In Africa, most of the social protection schemes are occupational and generally only cover people who have worked in the public sector, state enterprises, or large private firms. Many suggest African countries should adopt both formal and informal social security systems in order to maintain the traditional family support structures. There is no "one size fits all" model, and a range of different approaches has been adopted to meet the priorities based on varying national contexts.

Most of the social protection schemes in Africa were introduced after independence and are heavily influenced by their colonial heritage (Cohen and Menken 2006). In West Africa, countries such as Senegal, Ivory Coast, and Mali have all adopted social security programmes that are similar to the system used in France, which give defined benefit programmes to government workers based on their length of service and average earnings (Cohen and Menken 2006). In countries such as Ghana, Nigeria, and Zambia, there are provident funds, which are individual savings for workers from employers and employees for retirement funds. However, these funds are generally a once-off payment and are insufficient to support anyone in retirement. Armed conflict in many countries has widespread effects including destroying or stalling social security programmes in countries such as: Eritrea, Democratic Republic of Congo, Liberia, Sierra Leone, and Somalia (Cohen and Menken 2006). For example, one report suggests that, as a result of decades of conflict and mismanagement in the DRC, national institutions are in a state of turmoil which has meant that only a small minority of the elderly receive a pension (BBC 2010).

One assessment of social security systems in sub-Saharan Africa from 1960 to 1972 found that only seven countries had any form of a social security system that covered the elderly, namely: Burundi, DRC, Equatorial Guinea, Guinea, Mauritius, Rwanda, and South Africa (Kalasa 2001). By 1984 more countries had introduced social insurance for retired persons: Cameroon, Central African Republic, Congo, Ivory Coast, Benin, Ghana, Burkina Faso, Kenya, Lesotho, Liberia, Madagascar, Mali, Mauritania, Niger, Uganda, Tanzania, Togo, and Zambia (Kalasa 2001). However, the majority of these schemes only reached a small proportion of the older population, and was insufficient to meet basic needs. In addition, in some countries such as Uganda and Zimbabwe the manual processing of claims creates bureaucracy and delays in waiting to receive benefits which can deter people from seeking them (Cohen and Menken 2006).

To highlight an example from the above listed countries, in Ghana, old age is mainstreamed into the Growth and Poverty Reduction Strategy (GPRS II). This strategy includes a National Social Protection Strategy (2007) which began the Livelihood Employment Against Poverty (LEAP) and a Social Grants scheme which transfers cash to target groups to support basic needs (NPC 2007). The grant scheme:

> …will promote conditional and unconditional cash transfer systems and other support to displaced workers, while they seek employment, pregnant and lactating women, and provide target subsidies to the elderly, pensioners, smallholder farmers and people with disabilities. (Ghana Ministry of Environment, Science and Technology 2005)

Similar to other social security schemes, in Ghana, it only covers a small percentage of the working population. The national ageing policy recognises that the national economic and social structures are ill equipped currently to manage the increasing older population.

In Burkina Faso there is a social insurance trust and a system of mutual health insurance, however, the government recognises that the programmes do not reach the most vulnerable citizens including older people (HelpAge International 2008c). Mozambique established a cash transfer programme in the 1990s called the Gabinete de Apoio à População Vulnerável, which evolved into a programme now under the National Institute of Social Action, and supports identified groups of vulnerable people with a cash-based food subsidy (HelpAge International 2008c). However, this programme is only for urban areas, and there is an expressed need to expand to rural areas and a broader base of people. In Sierra Leone there is a National Social Security and Insurance Trust, and in 2005 a process began to develop a policy to provide support to older people. Tunisia has a rights-based social security system based on a contributory system linked to cash transfers and other benefits gained through employment (HelpAge International 2008c). Tunisia is currently exploring options of establishing universal social protection coverage based on age, for example. Algeria also offers older persons aged 65 years and older a pension, which was recently increased, however, there are calls to not only provide pensions but link them with health care for the elderly to ensure that they are given a good quality of life (Hadi 2010).

Currently there are only seven African countries that have put in place non-contributory pension systems for older persons: Botswana, Lesotho, Mauritius, Namibia, South Africa, Swaziland, and Tanzania (PensionWatch 2011). Lesotho's universal old-age pension is an example of a system where nutrition, education, and health access have improved, while the investments in local businesses have increased (HelpAge International 2008c). In Swaziland, an assessment of the old-age grant found that 71% of older people surveyed said that since receiving the non-contributory pension they were able to afford health care for themselves and 25% said they could now spend more on health care for other household members (RHVP 2010).

The South African pension programme was established in 1994. In 2002, pensions reached 1.9 million poor older people. Currently the pension covers 60% of the population that is eligible over 60 years of age, and uses 1.4% of the GDP (PensionWatch 2011). A study conducted in the province of KwaZulu-Natal found that many pensioners are supporting three generations with one pension (HelpAge International 2008b). However, one study showed that the old-age pension programme in South Africa has led to the improvement of the health and nutrition of girls, but has had no discernible effect on boys; this is due to the fact that pensions received by women had a large effect on child nutrition and health, while pensions received by men did not (Duflo 2003). Evidence from South Africa suggests that pensions can have positive effects on the health of members of the household, but it also depends greatly on how the cash transfer programmes are administered and who they are administered to (Duflo 2003). In addition, household survey data from Namibia found that cash transfers tend to reduce poverty, and that the effects are particularly positive for the poorest of the poor. Moreover, the transfers also tend to reduce inequality, but this impact is more limited (Levine et al. 2009).

With scarce resources, some argue that social pension funds are more effective if they target households with older people living with children and households with older people only. In 2003, Tanzania recognised in its National Ageing Policy that older people are among the poorest in society, often leading to unhealthy lifestyles, and in turn have limited access to health services. The National Ageing Policy makes provisions to establish a Revolving Loan Fund, with various stakeholders involved (the central government, local government, families, and voluntary agencies), to try and direct services to the informal sector, encourage savings and sensitise people to income generating activities as well as improve educational opportunities (Ministry of Labour, Youth Development and Sports 2003). Tanzania has also incorporated issues related to older persons in its National Strategy for Growth and Reduction of Poverty 2005–2010 (Sidorenko 2008). In addition, the government has started a Hunger Safety Net Program for people above the age of 55 (PensionWatch 2011).

Other countries have taken other economic measures, aside from pension funds, to target older populations with cash transfers. In Kenya, the Hunger Safety Net Programme (HSNP) is a government-led national social protection system to deliver cash transfers to poor and vulnerable people, which include older persons. Zambia's Pilot Kalomo Social Cash Transfer Scheme has been successful, and has broken myths that social cash transfer programmes are too expensive for developing countries and that poor people are often irresponsible with cash (HelpAge International 2008a). Part of the rationale for the establishment of the programme was that economically empowering caregivers can be an effective way to improve the welfare of orphans and vulnerable children, and provide them with their basic needs such as nutrition, education, and access to health services (Schubert 2005). Evidence from the scheme shows that almost half of the money was spent on investments in agriculture, with the other money going towards meeting basic needs such as school materials (HelpAge International 2008c). Some of the pay points were at designated rural health centres and schools, showing the link between health care and these welfare interventions.

Social security systems are helpful to empower older persons economically, however, they are not always successful in improving the health of the population and do not work in isolation. Often there is a low coverage of persons under the social security system, there are problems of corruption, and often the benefits are inadequate to sustain a reasonable standard of living.

11.8 Civil Society Responses to Ageing and Health

Civil society is starting to take a more active role in old-age-health-related activities and programmes. Often civil society fills a gap that governments do not cover, addressing pertinent issues related to population and development. Across various African countries, civil society groups have created interventions and programmes to target the ageing population and issues of older persons and health. Civil society also has a key role to inform and support policy development (HelpAge International 2008c).

In Uganda, the Uganda Reach the Aged Association (URAA) was founded in 1991. Changing socio-economic conditions in the country caused a shift in traditional family structures leaving older persons without care, in addition to placing an increased burden of caring for orphaned grandchildren as well as increased poverty (URAA 2011). The URAA is involved in training and supporting community volunteers which come together to form Village Health Teams that offer home-based care and support to older persons, specifically targeting those that are head of households and affected by HIV/AIDS. URAA also aims to provide psychosocial support to older persons, and train older people as volunteers, peer educators, and counsellors. URAA works in collaboration with the Ministry of Gender, Labor, and Social Development, and other Ministries to improve the livelihoods of older people (URAA 2011). This organisation arose in direct response to a need that was neglected by the government and it continues to attempt to fill this gap around the country.

A similar organisation to URAA is Age-in-Action, a South African NGO targeting older people. Age-in-Action was established in 1956, providing services nationwide, focusing on community-based health care for older persons, to initiate prevention and educational programmes on HIV/AIDS for older persons. It is an umbrella organisation that has over 800 NGOs as members who provide services to older persons in need of care (Age-in-Action 2011). Age-in-Action also strives to create community initiatives to develop community groups for older person's support. In Nigeria, concern about the socio-economic rights and needs of the elderly has become the focus of attention of a NGO known as Ageing with Grace and Dignity (AGRAD). Their work is primarily in the form of advocacy, to ensure that governments at all levels in the country integrate ageing into their development plans. The organisation subscribes to the belief of active ageing, arguing for the continuing participation of the elderly in social, economic, cultural, spiritual, and civic activities.

During the 2008 African Union Commission meetings, the President of Cameroon emphasised the role of civil society in supporting and informing the development of social protection systems regionally and nationally. President Paul Biya was quoted as saying:

> Development is not an end in itself. In a democracy like ours, which puts the human person at the centre of the system, development must primarily drive the improvement of human life (HelpAge International 2008c).

The Regional Centre for the Welfare of Ageing Persons in Cameroon was established by Prince Bengha Martin in an effort to lobby on behalf of senior citizens to create income-generating initiatives for elderly people, like pig breeding and mushroom farming (Martin 2008). Similar to many African countries, Cameroon does not have any government institutions, organisations, or health care services that specialise in meeting older people's needs, so this organisation seeks to create income to help older persons provide for their own needs as well as a surplus for the needs of others.

The international NGO HelpAge International (mentioned earlier), works in ten African countries in a variety of capacities: Burkina Faso, Ethiopia, Ghana, Kenya, Mozambique, South Africa, Sudan, Tanzania, Uganda, and Zimbabwe. In Kenya, it has helped train more than 300 people to become HIV/AIDS community educators. Similarly, in Tanzania, it has conducted workshops to give older people the skills and confidence to take care of relatives living with HIV/AIDS. In Ethiopia, it has designed programmes to train 300 burial service providers to share their knowledge of the impact of AIDS on older people to the surrounding communities. And in South Africa it has worked with traditional healers on creating HIV prevention efforts.

HelpAge International partners with local NGOs and governments acting as an informer and supporter of national policies and programmes. For example, in Burkina Faso, it works with partners to empower older women to realise their rights, and provides emotional support to women who are refugees as a result of witchcraft accusations along with the Ministry of Social Action and National Solidarity. In

Sudan, HelpAge International works with older people in refugee camps who have fled West Darfur conflict by providing them with health care, eye care, literacy training, and social activities.

The advocacy work done by civil society is crucial to influence national policies and raise awareness of older people's rights, empowering them to help them get health care, education, legal help, and financial support, which in turn creates stronger communities and societies at large. The involvement of older people in the process is of high priority.

11.9 Conclusion

Population ageing that is occurring now is unprecedented, and over the next decade the number of people aged 60 years and over is expected to rise substantially, and most of the increase will occur in developing countries. Globally, the number of older persons is expected for the first time to exceed the number of children in 2045 (United Nations 2009). The rapidly ageing population is likely to have far-reaching effects. Economically, population ageing affects growth, savings, investments, consumption, labour markets, social security systems, taxation, and intergenerational transfers; socially, it influences family structures, housing demands, migration trends, and the epidemiology and need for health care services; and politically, it shapes voting patterns and representation (United Nations 2009). Population ageing has received considerable attention in developed countries, but only in the past few decades has awareness of the ramifications of a rapidly ageing population grown in the rest of the world.

Until relatively recently, population ageing in Africa has not received much attention largely because of the continent's youthful population. However, mounting evidence suggests that the number of older people in the population is steadily increasing and is likely to multiply over the next few decades. Ageing in African countries can be attributed almost exclusively to declines in fertility and mortality rather than a result of improved socio-economic development, as seen in many developed countries (UNFPA 2002). Even in the context of a major and devastating AIDS pandemic, the older population is expected to steadily rise in Africa (Bongaarts 2005). In addition to the increase in the share of older people in the African population, a gender disparity in gains in life expectancy is likely to emerge, resulting in women outnumbering men at the older age groups. The implications of an ageing population and the feminisation of the older population are major issues with which policy makers must grapple.

The World Assembly on Ageing held in Madrid was instrumental in encouraging African governments to focus more attention on the issue of ageing. The International Plan of Action emphasises the crucial role of governments in "promoting, providing and ensuring access to basic social services, bearing in mind specific needs of older persons" (United Nations 2002). The Madrid Plan also made it clear that implementation of policy was the responsibility of the national government. According to the

Madrid Plan of Action, population ageing is strongly related to national development agendas, because most developing countries will need to face development challenges in parallel with rapid population ageing, while developed countries experienced a more gradual demographic transition (United Nations 2002). Moreover, the policy challenges will be different for most developing countries, as the majority of older persons will be living in multi-generational households, and in rural areas (United Nations 2002). Given the important role of caregiving that many older people adopt in their households, it is particularly important that they are given the opportunity to remain economically and socially active in order to ensure their health and well-being, a call strongly articulated in the Plan of Action.

The implementation of the Madrid Plan of Action varies across countries in Africa. Some countries have made progress in formulating policies on ageing and have adopted national policies on ageing, while others have not even begun. In Cameroon, budgetary constraints have been blamed for the lack of action in the formulation of policies. In Mozambique, the problem has been attributed to the failure to allocate an overall budget for the National Plan (UNFPA and HelpAge International 2011). In resource-poor countries, the implementation of plans of action on ageing is hampered by human and financial constraints. In a highly competitive funding environment, other development initiatives are often given first preference. The problem is further compounded by the lack of accurate and reliable data and information on ageing, especially in conflict and emergency situations. Population ageing needs to be a high-priority agenda of national governments in Africa, and this should also be reflected in national budgets. Older people should be directly involved in the design and implementation of policies that affect their health and well-being.

In much of Africa, the family remains the most important source of support for older people. Older people in Africa are highly valued and occupy an elevated status in many communities. However, due to socio-economic and political changes, and increasing conflict and instability, older people are increasingly becoming marginalised and are often seen as a burden. At the same time, older people increasingly find themselves taking on the responsibility for the care of their sick children, and also orphaned grandchildren. Africa is suffering from a devastating AIDS pandemic. Almost two-thirds of people living with AIDS are found in sub-Saharan Africa. Population forecasts offer several possible scenarios with respect to the impact of HIV/AIDS on the demographic transition. Perhaps one of the leading assumptions is that the pandemic will result in increased mortality coupled with a slight decline in fertility rates. The importance of understanding the effects of HIV/AIDS on all population age groups notwithstanding, there remains a relative paucity of research on the impact of the pandemic on the older population. A priority is, therefore, to intensify efforts to address the issue of HIV/AIDS and its impact on different age groups in the population, particularly in African countries that are suffering the worst effects of the pandemic.

Older people, particularly women, suffer from discrimination due to age. Policies and interventions should focus more specifically on improving the situation of older people in Africa. Older women have been identified as particularly vulnerable. They are more likely to suffer physical abuse and violence due to negative societal attitudes and harmful traditional and customary practices. Exclusion from ownership of land, credit, and income-generating programmes contributes to their loss of independence, status, and dignity. In addition, they are more likely to bear the brunt of the HIV/AIDS pandemic in terms of productive time devoted to caregiving and the distribution of financial resources to other members of the household (United Nations 2002). As such, interventions to develop skills training as a component of caregiving would likely go a long way towards both engaging with older women, and providing critical support and skills' development to households burdened with the effects of HIV/AIDS. Other specific interventions targeting this group should include: improved access to land rights, credit markets, social security services, and health care services (United Nations 2002). The protection of the rights of older people is enshrined in a number of international conventions, but more emphasis should be placed on protecting them against human rights violations, especially in emergency situations and political unrest.

Policies and programmes directed at older persons not only benefit individuals, but also households, communities, and the economy. The contribution of cash transfers in the form of the old-age pensions is just one example. A number of African countries have introduced the old-age pension scheme to ease the economic hardships experienced by the poorest older population. Increasing evidence suggests that the benefits of the old-age pension extend beyond the individual. Barrientos (2006) observes that cash transfers, especially if targeted to poor households, tend to have an impact on the younger generation, especially where the elderly live in multi-generational households, as benefits are shared with other members and, as a result, leaves the older person less likely to feel burdensome to their families. In South Africa, the impact of the old-age pension has resulted in improved health, nutrition, and better educational outcomes for household members (Case 2001; Duflo 2003). Older men and women previously employed in the formal sector may receive pensions from contributory social security schemes, but the vast majority of older populations across Africa involved in the informal sector are not covered by these schemes and therefore do not receive any financial support. Extending the coverage of social security systems is likely to have wider benefits for society. However, the challenge is to improve governance in many countries in Africa and build credibility of social security institutions (Barbone and Sanchez 1999). Poor administrative structures, corruption, limited resources, and political instability have undermined social security systems in many African countries (Charlton and McKinnon 2001).

One of the most immediate challenges in the coming years is meeting the health needs of an older population. As the number of older people grows, and people live longer, the demand and cost of health care is likely to rise dramatically, due to the heavy burden of chronic conditions. Chronic conditions, as defined by WHO, require ongoing management over a number of years or decades. This may require massive restructuring of health systems in order to expand services. At present, in most African countries, health systems are not geared towards managing chronic conditions, especially for the aged. Chronic conditions may impose both direct and indirect costs, all of which serve to constrain health care access and influence the decision to seek treatment. The direct costs include the expense of drugs, treatment, and care. While the cost of drugs often contributes a sizeable share of direct costs, transport costs and often overlooked expenses can also be considerable. The indirect costs include lost productive time for sick and other household members. Many older people do not seek care when ill as a result of these barriers to access. Improving access to health care services in Africa may have long-term benefits. Tackling ill health will assist in the reduction of poverty by enhancing quality of life, expanding opportunities, and safeguarding livelihoods and productivity. Nonetheless, the poor are most vulnerable to ill health, and have the least means to address it. This reality, coupled with limited access to affordable medicines and inadequate health systems to deliver them, in many African countries, has meant that significant inequities continue to exist. Government intervention in this regard is therefore crucial and policies have to focus more attention on older people who are most likely to have an increased need for health care.

References

Aboderin, I. (2006). African Women and Ageing: Nairobi, Beijing, and the implications for African gender scholarship. *CODESRIA Bulletin, 1&2*, 23–25.

Aboderin, I. (2008). *Advancing health service provision for older persons and age-related non-communicable disease in sub-Saharan Africa: Identifying key information and training needs. African Research on Ageing Network (AFRAN) policy-research dialogue*. Abuja: AFRAN.

Aboderin, I. (2010). Understanding and advancing the health of older populations in sub-Saharan Africa: policy perspectives and evidence needs. *Public Health Reviews, 32*, 1–14.

Age-in-Action. (2011). Age-in-action. http://www.age-in-action.co.za. Accessed 08 May 2011.

Asagba, A. (2005). Research and the formulation and implementation of ageing policy in Africa: the case of Nigeria. *British Society for Gerontology Generations Review, 15*(2), 39–41.

Barbone, L., & Sanchez, L. A. (1999). Pension and social security in Sub-Saharan Africa: Issues and options. Paper presented at the XIII International Social Security Association African Regional Conference, Accra, Ghana, 6–9 July.

Barrientos, A. (2006). Poverty reduction: the missing piece of pension reform in Latin America. *Social Policy and Administration, 40*(4), 369–384.

Bashir, S., Salih, O. A., & Magied, A. A. (2007). Nutrition status, health risk factors and food security of older people living in Kass Province case study: Shattaya Fur. *Ahfad Journal: Women and Change, 24*(2), 21–40.

BBC News. (13 September 2010). Lost pension hopes in DR Congo. http://www.bbc.co.uk/news/world-africa-11288154?print=true. Accessed 05 Jan 2011.

Bongaarts, J. (2005). *Demographic trends*. Los Altos, CA: David and Lucile Packard Foundation Population Program Review Task Force.

Botswana Ministry of Health. (n.d.). Older population and health system: A profile of Botswana. Botswana: Ministry of Health.

Bremner, B., Frost, A., Haub, C., Mather, M., Ringheim, K., & Zuehlke, E. (2010). World population highlights: key findings from PRB's 2010 world population data sheet. *Population Reference Bureau, 65*(2), 1–12.

Case, A. (2001). *Does money protect health status? Evidence from South African Pensions. NBER Working Paper 8495*. Cambridge, MA: National Bureau of Economic Research.

Castelo, M. A., Gaspan, M., & Felix, B. V. (1999). A cultural approach to HIV/AIDS prevention and care: Angola's experience. UNESCO/UNAIDS Research Project. http://unesdoc.unesco.org/images/0012/001206/120612e.pdf. Accessed 15 Jan 2011.

Charlton, R., & McKinnon, R. (2001). *Pensions in development*. London: Ashgate.

CNCD. (2011). Consortium for non-communicable diseases prevention and control in sub-Saharan Africa: Introduction. http://www.aihdint.org/cncd/information.php?MenuID=10&SubMenuID=7. Accessed 20 Feb 2011.

Cohen, B., & Menken, J. (Eds.). (2006). *Aging in Sub-Saharan Africa: Recommendations for furthering research*. Washington DC: The National Academies Press.

Coumé, M. (2009). *Rapport Sur le Plan Sesame de Soins Gratuits Pour les Personnes Agees de 60 an et Plus au Senegal: Revue 2008:/Meridien President*. Dakar: Ministry of Health.

Duflo, E. (2003). Grandmothers and granddaughters: old age pensions and intrahousehold allocation in South Africa. *The World Bank Economic Review, 17*(1), 1–25.

Dullemen, C. (2006). Older people in Africa: new engines to society? *Feminist Formations, 18*(1), 99–105.

Ewing, D. (1999). Gender and ageing. In J. Randel, T. German, & D. Ewing (Eds.), *The ageing and development report: Poverty, independence and the world's older people*. London: Earthscan.

Ghana Ministry of Environment, Science and Technology. (2005). *Growth and poverty reduction strategy (GPRS II)*. Accra, Ghana: Ministry of Environment, Science and Technology.

Ghana Ministry of Manpower, Youth and Employment. (2007). *The National Social Protection Strategy (NSPS): Investing in People. Accra*, Ghana: Ministry of Manpower, Youth and Employment.

Giles, W. (2010). Preventing non-communicable diseases in sub-Saharan Africa. *Global Health Promotion, 17*, 3–5.

Hadi, H. E. (2010). *Rise in elderly allowance stirs debate in Algeria*. Algiers: Magharebia.

HelpAge International. (2003). *African union policy framework and plan of action on ageing*. Nairobi: African Union.

HelpAge International. (2008a). Africa makes progress on social protection. *Ageing in Africa, 31*, 1–8.

HelpAge International. (2008b). Older people in Africa: A forgotten generation. http://www.helpage.org. Accessed 20 May 2011.

HelpAge International. (2008c). *Investing in social protection in Africa: Summary report of national consultations held in Burkina Faso, Cameroon, Mozambique, Rwanda, Sierra Leone and Tunisia*. London: HelpAge International.

HelpAge International. (2010). *Discrimination against older women in Burkino Faso*. London: HelpAge International.

Joubert, J., & Bradshaw, D. (2006). *Population ageing and health challenges in South Africa* (Chronic diseases of lifestyle in South Africa 1995–2005). Tygerberg: Medical Research Council.

Kalache, A. (2008). The World Health Organization and global aging. In M. Robinson (Ed.), *Global health and global aging*. San Francisco: Jossey-Bass.

Kalache, A., & Sen, K. (1999). Ageing and health. In J. Randel, T. German, & D. Ewing (Eds.), *The ageing and development report: Poverty, independence and the world's older people*. London: Earthscan.

Kalasa, B. (2001). *Population and ageing in Africa: A policy dilemma?* Addis Ababa: UNFPA.

Levine, S., van der Berg, S., & Yu, D. (2009). *Measuring the impact of social cash transfers on poverty and inequality in Namibia*. Cape Town: University of Stellenbosch.

Lloyd-Sherlock, P. (2000). Population ageing in developed and developing regions: implications for health policy. *Social Science & Medicine, 51*, 887–895.

Lloyd-Sherlock, P. (2002). *Conference news: Ageing, development and social protection*. Madrid: UNRISD.

Martin, B. (2008). *Helping hand for Cameroon's elderly*. London: Guardian Weekly.

Mba, C. (2010). Population ageing in Ghana: research gaps and the way forward. *Journal of Aging Research I, 2010*, 1–8.

McLigeyo, S. O. (2002). Ageing population in Africa and other developing communities: a public health challenge calling for urgent solutions. *East African Medical Journal, 79*(6), 281–283.

Ministry of Health. (2005). *Reversing the trends: The second National Health Sector Strategic Plan of Kenya (NHSSP II 2005–2010)*. Nairobi: Ministry of Healthi.

Nigeria Federal Ministry of Health. (2004). *Revised National Health Policy*. http://www.herfon.org/docs/Nigeria_NationalHealthPolicy_sept_2004.pdf. Accessed 11 Feb 2011.

NPC. (2007). *Ghana country report on the implementation of the Madrid International Plan of Action on Ageing (MIPAA)*. Accra: National Population Council.

Oxford Business Group. (2008). *The Report: Senegal 2008*. http://www.oxfordbusinessgroup.com/product/report-senegal-2008. Accessed 1 Feb 2011.

Padarath, A., Searle, C., & Esu-Williams, E. (2006). *Understanding barriers to community participation in HIV and AIDS services. Final report*. Johannesburg: Population Council.

PensionWatch. (2011). Social protection in older age. http://www.pension-watch.net. Accessed 20 Feb 2011.

Randel, J., German, T., & Ewing, D. (Eds.). (1999). *The ageing and development report: Poverty independence and the world's older people*. London: Earthscan Publications Limited.

RHVP. (2010). *Swaziland old age grant impact assessment*. London: HelpAge International.

Schubert, B. (2005). *The pilot social cash transfer scheme Kalomo district—Zambia. Chronic Poverty Research Center Working Paper*. Manchester: Chronic Poverty Research Center.

Sidorenko, A. (2008). World policies on aging and the United Nations. In M. Robinson (Ed.), *Global health and global aging*. San Francisco: Jossey-Bass.

Sunkur, J. (2007). *Republic of Mauritius: National report on the review and appraisal of the implementation of the MIPAA*. Addis Ababa: Expert Group Meeting on Ageing in Africa.

Suzman, R. (2010). Guest Editorial: The INDEPTH WHO-SAGE multicentre study on ageing, health and well-being among people aged 50 years and over in eight countries in Africa and Asia. *Global Health Action* 3, 5–7.

Tanzania Ministry of Labour, Youth Development and Sports. (2003). *National Ageing Policy*. http://www.tanzania.go.tz/pdf/NATIONAL%20AGEING%20POLICY%20-%20FINAL%20PAPER.pdf. Accessed 14 Jan 2011.

United Nations. (1948). The Universal Declaration of Human Rights. http://www.un.org/en/documents/udhr/index.shtml. Accessed 14 Feb 2011.

United Nations. (1991). United Nations principles for older persons. http://www.un.org/ageing/un_principles.html#Principles. Accessed 13 Feb 2011.

United Nations. (1996). Standard Rules on the Equalization of Opportunities for Persons with Disabilities. http://www.un.org/documents/ga/res/48/a48r096.htm. Accessed 13 Feb 2011.

United Nations. (2002). *Madrid International Plan of Action on Ageing, 2002. Second World Assembly on Ageing*. New York: United Nations Economic and Social Affairs.

United Nations. (2009). *World population ageing 2009*. New York: United Nations Department of Economic and Social Affairs.

UNFPA. (2002). *Population ageing and development: Social, health and gender issues*. Malta: UNFPA.

UNFPA, & HelpAge International. (2011). *Overview of available policies and legislation, data and research, and institutional arrangements relating to older persons—progress since Madrid*. New York: UNFPA and HelpAge International.

URAA. (2011). Uganda Reach the aged association: dignified self-fulfilled poverty free ageing Uganda. 2011. http://uraauganda.org/services.html. Accessed 20 Jan 2011.

WHO. (2002). *Active ageing: A policy framework. Second United Nations World Assembly on Ageing*. Madrid: WHO.

WHO. (2011a). Ageing. 2011. http://www.who.int/topics/ageing/en/. Accessed 21 Jan 2011.

WHO. (2011b). Definition of an older or elderly person: proposed working definition of an older person in Africa for the MDS project. http://www.who.int/healthinfo/survey/ageingdefnolder/en/index.html. Accessed 16 Jan 2011.

WHO. (2011c). Poverty and health. 2011. http://www.who.int/hdp/poverty/en/. Accessed 16 Jan 2011.

Index

P. Maharaj (ed.), *Aging and Health in Africa*, International Perspectives on Aging 4,
DOI 10.1007/978-1-4419-8357-2, © Springer Science+Business Media New York 2013

C

D

MIX
Papier aus verantwortungsvollen Quellen
Paper from responsible sources
FSC® C105338

If you have any concerns about our products,
you can contact us on
ProductSafety@springernature.com

In case Publisher is established outside the EU,
the EU authorized representative is:
Springer Nature Customer Service Center GmbH
Europaplatz 3, 69115 Heidelberg, Germany

Printed by Libri Plureos GmbH
in Hamburg, Germany